SpringerBriefs in Electrical and Computer Engineering

More information about this series at http://www.springer.com/series/10059

Osameh M. Al-Kofahi • Ahmed E. Kamal

Resilient Wireless Sensor Networks

The Case of Network Coding

Springer

Osameh M. Al-Kofahi
Yarmouk University
Irbid, Jordan

Ahmed E. Kamal
Iowa State University
Ames, IA, USA

ISSN 2191-8112 ISSN 2191-8120 (electronic)
SpringerBriefs in Electrical and Computer Engineering
ISBN 978-3-319-23963-7 ISBN 978-3-319-23965-1 (eBook)
DOI 10.1007/978-3-319-23965-1

Library of Congress Control Number: 2015953219

Springer Cham Heidelberg New York Dordrecht London

Printed on acid-free paper

Springer International Publishing AG Switzerland is part of Springer Science+Business Media (www.springer.com)

To my parents and my wife.
  O. A.

To the memory of my parents.
  A. K.

Preface

Wireless Sensor Networks (WSN) have found many applications in many domains. However, due to the deployment of WSNs in harsh environments, and also due to the battery powered nature of sensors, sensors may fail or cease to function properly. This Brief considers the resilient operation of WSNs in which data delivery from sensor nodes to the sink is guaranteed, even when some sensors fail, or when communication is impaired. Traditional protection schemes are either slow in reacting and recovering from failures or expensive in terms of backup resources, which are reserved to recover from failures. Therefore, this Brief covers protection strategies that use the technique of network coding, which have the advantages of overcoming the deficiencies of the traditional schemes.

Since network coding-based protection can be implemented in several ways, this Brief is divided into a number of chapters addressing theoretical bases and practical implementations. After a brief introduction to WSNs and the resilience problem in WSNs, this Brief introduces a basic centralized scheme, which applies to a restricted network topology, and will be introduced as an optimal solution. Then, the topology is relaxed to include practical wireless network topologies, and the scheme is generalized to apply to such topologies. Coding algorithms will be covered, which are based on using coding coefficients from the binary field. These include algorithms for performing deterministic coding with $\{0,1\}$ coefficients at intermediate sensor nodes and for decoding at the sink to recover data. For large-scale WSNs, this Brief covers distributed network coding approaches, which can be used to recover from failures and from packet losses in WSNs. The transmission scheduling problem is important in order to achieve efficient usage of the wireless spectrum and avoid transmission collisions. Therefore, this Brief covers scheduling algorithms when digital network coding and analog network coding are employed. Practical considerations for the implementation of those algorithms will be covered.

This concise but in-depth coverage of network coding-based protection of WSNs should be of interest to researchers and graduate students in Electrical and Computer Engineering, as well as in Computer Science, who are interested in the topic of resilient WSNs. It should be also of interest to professionals working in the deployment of resilient WSNs.

The authors would like to thank graduate students and faculty members in the Department of Electrical and Computer Engineering at Iowa State University for numerous discussions and comments about the research that led to this Brief.

Irbid, Jordan   Osameh M. Al-Kofahi
Ames, IA, USA   Ahmed E. Kamal
July 2015

Contents

1 Introduction 1
    1.1 Wireless Sensor Networks 1
    1.2 Wireless Sensor Networks Failures 2
    1.3 Resilient Operation of WSNs 3
    1.4 Network Coding: Concepts and Advantages 5
    1.5 Book Organization 7
    References 7

2 Network Coding-Based Resilient WSNs: The Centralized Approach .. 9
    2.1 Problem Description 9
    2.2 Network Coding-Based Resilience Approach 11
        2.2.1 Assumptions, Definitions and Notation 11
        2.2.2 Sufficient and Necessary Conditions 13
    2.3 Generalizations and Practical Considerations 15
        2.3.1 General Network Topology 15
        2.3.2 Multiple Failures 16
        2.3.3 The Case of $|L| > n + 1$ 16
        2.3.4 Networks with Limited Minimum Cuts 17
        2.3.5 MILP Formulation 18
    2.4 Coding 20
        2.4.1 Path Coding and Tree Coding 20
        2.4.2 Constructing a Coding Tree 23
    2.5 Network Performance 25
    2.6 Chapter Summary 28
    References 28

3 Network Coding-Based Resilient WSNs: The Distributed Approach ... 29
    3.1 Tolerating a Single Loss 29
        3.1.1 Assumptions and Notation 29
        3.1.2 Rules of Operation 30
    3.2 Tolerating Multiple Losses 34

3.3   Coding/Decoding Issues 36
      3.3.1   Relative Indexing for Efficient Encoding 36
      3.3.2   Best Effort Decoding 38
3.4   Routing for Maximally Disjoint Paths 40
3.5   Selecting Parameter h 44
      3.5.1   Evaluating $P(rcv)$ 47
3.6   Simulation Results 49
3.7   Chapter Summary 51
Reference 52

4 Transmissions Scheduling in Network Coding-Based
   Resilient WSNs 53
4.1   Introduction 53
4.2   Scheduling Based on Digital Network Coding 55
4.3   Scheduling Based on Analog Network Coding 57
      4.3.1   Special Case: When $\Delta_L = 2$ 58
      4.3.2   Special Case: When G is a Tree 59
      4.3.3   Maximum Gain of ANC-Based Scheduling 61
4.4   Performance Evaluation and Comparison 62
4.5   Chapter Summary 64
References 65

5 Conclusions 67

Chapter 1
Introduction

This chapter provides a brief background to wireless sensor networks (WSNs) and the resilience problem in WSNs. Standard approaches to providing resilience in WSNs will be discussed, followed by an introduction to the technique of network coding that is the basis of the resilience approaches introduced in this brief.

1.1 Wireless Sensor Networks

During the past fifteen years, significant research and development activities have taken place in the area of wireless sensor networks (WSNs) [1, 2]. WSNs have found applications in many domains including surveillance, precision agriculture, public safety, environmental, fire monitoring, health, smart homes, military applications, and many other domains. The availability of low cost new hardware, as well as the promise of new functionalities have contributed to this popularity. The Internet of Things (IoT) is an outgrowth of WSNs and many of the concepts developed for WSNs have been adopted and extended to IoT.

A WSN consists of sensor nodes (see Fig. 1.1) which are equipped with sensors that sense one or more phenomena, e.g., temperature, humidity, light, moisture, etc., depending on the application. A sensor node also has a radio chipset allowing wireless communications between nodes, which usually uses radio frequency (RF) signals, although other signals, like ultrasound and visible light signals, have been used in some application domains. These components are controlled by a micro-controller, typically with limited computing power and storage space. Most sensing devices provide measurements in an analog format, and therefore interface to the micro-controller using analog-to-digital convertors (ADC), while some provide the readings in digital format, e.g., GPS chips. All of these components

© The Author(s) 2015
O.M. Al-Kofahi, A.E. Kamal, *Resilient Wireless Sensor Networks*, SpringerBriefs in Electrical and Computer Engineering, DOI 10.1007/978-3-319-23965-1_1

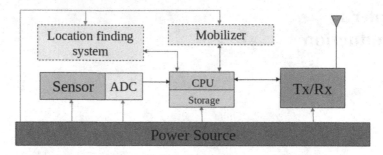

Fig. 1.1 Sensor node

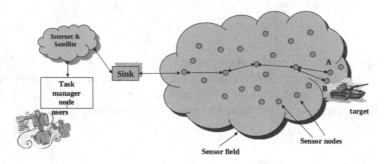

Fig. 1.2 A wireless sensor network

are powered by a power source, typically a battery, with the option of using energy harvesting to replenish the battery energy without the need to recharge it.

Sensor nodes in a WSN sample the environment, and this can be either event-driven or periodic. The sampled measurements are sent to a central sink node (see Fig. 1.2) which collects all the measurements, analyzes them, and may then take or recommend actions, e.g., water or fertilize a farming field in a precision agriculture application. Since in many applications sensors are deployed over a large area, and it is infeasible to communicate with the central node in a single hop, multi-hop ad-hoc communication is used in which sensor nodes receive data packets from other sensors and relay these data packets to other nodes, until they are eventually delivered to the central node. Moreover, since communication converges toward to the central node, this communication is usually referred to as *many-to-one* communications or *convergecast* communications.

1.2 Wireless Sensor Networks Failures

WSNs are prone to failure, which can be due to many causes that include the following:

- Sensor nodes are usually powered by batteries, and the energy of such batteries may be depleted. The battery energy depletion time depends on several factors such as the battery capacity, the sampling frequency, the range of communication, the amount of computations, etc. If batteries are not replenished in time, and if energy harvesting is not used, sensor nodes may run out of power and may fail. It is not only that the sensor node will stop functioning and produce measurements, but they may also disrupt communications if they are used to relay readings from other sensor nodes to the central node.
- Sensor nodes may be used in harsh environments, like sensors used in military applications or sensors used by firefighters. In this case the sensor nodes themselves may be physically damaged and will stop functioning.
- Sensor nodes may not fail, but the communication channels might be impaired. For example, physical barriers absorb RF signals, and the quality of the signal may be severely degraded if the signal propagates through multiple barriers. It was also shown that increased temperature can have adverse effects on the quality of the signal and may lead to signal impairments. Therefore, communication among sensors used in fire environments may suffer from this type of impairment. Malicious jamming of the channel, e.g., by enemy combatants in military applications, may cause the signal to be highly impaired.
- When sensor nodes change location, either intentionally when sensor nodes are equipped with mobilizers, or inadvertently when sensor nodes move due to a physical event like mud slide, or structure collapse. In this case movement and location changes may result in disrupting communications due to making some sensor nodes out of communication range and therefore unreachable.

In some cases, like channel impairments, automatic repeat request (ARQ) and forward error correction (FEC) strategies can be used to recover affected data. However, if the bit error rate (BER) is too high, such strategies will fail to recover the data. In some other cases, like physical damage, ARQ and FEC will not be useful.

Providing resilient operation of WSNs has been a subject of interest due to the wide use and deployment of WSNs, and the requirement that such networks function properly, even in the presence of sensor node or channel failures.

1.3 Resilient Operation of WSNs

Network resilience in general is defined as the capability of the network to transmit and relay data successfully in a timely manner, even in the presence of failures. Network resilience is important to sustain continuous uninterrupted service for network users and for the application. In [3] the authors provide a survey of survivability mechanisms for multihop wireless networks in general, which include WSNs. More recently, Younis et al. [4] provides a comprehensive survey of topology management techniques for providing resilience against node failures in WSNs.

In general, mechanisms for achieving network resilience can be divided into the following categories:

- **Protection mechanisms:** Protection is usually achieved by using redundant network resources to carry redundant data units. Usually, a data unit is duplicated and forwarded on multiple paths from the source to the destination. In this case, a data delivery failure occurs and will be detected only if all paths fail. Otherwise, there is no need to detect the failure or retransmit the information. This is called proactive protection and is usually referred to as 1+1 protection. An alternative way to provide protection is to divide the paths into two sets, primary and backup, where only the primary path is used to forward data to the destination. A backup path will only be used if the primary path fails. This called reactive protection and is usually referred to as 1:1. Reactive protection can be extended to M:N, where M backup paths are reserved to protect N primary paths. The M backup paths are shared by the N sources, and can be used by any source if a failure occurs on its primary path, which makes this type of protection more efficient in utilizing the network resources. However, reactive protection is slower than proactive protection since a source must detect a failure first, and then switch the data flow to one of the available backup paths. Although reactive protection is known and used in wired networks, it is not technically accurate to talk about path reservation in wireless networks, since there are no actual physical links that can be reserved. However, a node in a wireless network might learn multiple paths to the destination during the route discovery process and can use them in a fashion similar to that of reactive protection and without reserving the paths in advance.

- **Restoration mechanisms:** In restoration mechanisms only a single path is used from a source to a destination, and no backup paths are found in advance. Therefore, restoration mechanisms consume fewer resources than protection mechanisms. However, restoration does not provide recovery at the speed of protection since failures need to be detected first, and then a resource discovery procedure is invoked. Then, rerouting is done to find a different route for the data units. Note that the rerouting mechanism here is different from that in reactive protection. In restoration, no information about the available network resource is known to the node that detected the failure, that is why it needs to discover the network resources first to be able to do the rerouting afterwards. However, under protection, multiple paths are computed a priori, and thus, the rerouting mechanism in reactive protection is very simple and is confined to just switching to an available path from the backup set. It should be noted that restoration is implicitly implemented in all routing protocols in the form of route maintenance mechanisms.

- **Hybrid mechanisms:** In this case a mix of protection and restoration mechanisms can be used together.

Designing network resilience mechanisms has been centered around achieving two objectives. The first is to minimize the amount of network resources used for recovery from failures, e.g., backup link capacities. The second is an agile recovery operation, i.e., reroute data over the backup paths with the least delay. These two objectives are usually contradictory. Minimizing network resources requires sharing of backup resources between connections, and if this is employed then upon the occurrence of a failure, the failure needs to be detected and attributed to the affected connections. Then, operation at the intermediate nodes needs to be altered to support rerouting of affected data. These operations add to the delay in recovering from failures. However, if an agile failure recovery strategy is used, e.g., 1+1, then more than 100% of the network resources need to be added as backup resources.

Jointly achieving the above two objectives has been a long sought objective of network operators and protocol designers, and several studies have been conducted to achieve this. The technique of network coding, which will be introduced in the next section, has made it possible to jointly achieve the two objectives.

1.4 Network Coding: Concepts and Advantages

Network coding was introduced in [5]. Rather than forwarding packets at intermediate nodes, which is the standard operation under routing, network coding allows the intermediate nodes in a network to form combinations from packets arriving at their input ports and then forward these combinations on their output ports. Typically, these combinations are linear and they are performed by using coefficients from a certain field and are multiplied by the incoming packets. These products are summed up, also over the same field. It was shown in [5] that by using network coding in a multicast connection, the multicast capacity, which is defined as the smallest of the minimum cuts between the multicast source and each of the destinations, can be achieved.

As an example, consider the network shown in Fig. 1.3.(a). In the figure, there is one multicast session in which node A is a multicast source that transmits the packets b_1 and b_2. Nodes E and G are two destinations of the same multicast session, and they should both receive the same two packets b_1 and b_2. Without network coding link (C, F) will be the bottleneck link and will have to alternate between

Fig. 1.3 An example of multicasting: **(a)** without network coding; **(b)** with network coding

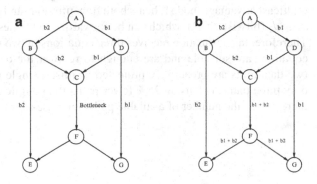

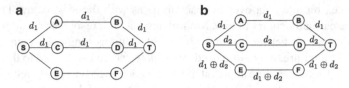

Fig. 1.4 Proactive protection: **(a)** without network coding; **(b)** with network coding

sending b_1 and b_2. If network coding is employed, as shown in Fig. 1.3.(b), node C can combine b_1 and b_2. Under network coding, the multicast capacity will be achieved and link (C, F) will not be a bottleneck. This is illustrated in Fig. 1.3.(b) where the addition here is modulo 2 (i.e., equivalent to bitwise XOR). In this case node E recovers b_1 by XORing b_2 with $b_1 + b_2$, and similarly G recovers b_2 by XORing b_1 with $b_1 + b_2$.

The theory of network coding was further investigated afterwards, and it was shown that linear network coding is sufficient to achieve the multicast capacity in [6]. A combinatorial approach for decomposing a multicast network to identify where the combinations performed in the network was introduced in [7], and polynomial time algorithms were proposed in [8] to assign coding vectors to network links, i.e., to define the relationship between the symbol carried on a link with the set of symbols arriving at the tail node of that link. Many proposals for employing network coding in a variety of applications have been introduced, and interested readers may refer to [9] for a comprehensive review of such applications.

Our focus in this brief is on applying network coding in order to provide resilient operation in WSNs. We therefore conclude this chapter by introducing a simple example to show the advantage of using network coding, rather than alternate routing, in order to implement proactive protection. Consider the example in Fig. 1.4a, where there are three paths between S and T. In this case, using duplication (i.e., sending two copies of the same data unit to the destination) we cannot protect more than one data unit, since to protect two data units we need four disjoint paths. However, if network coding is allowed, as shown in Fig. 1.4b, the source can send two different (not duplicates) data units to two neighboring nodes on two disjoint paths; d_1 to node A and d_2 to node C. Because of the wireless multicast advantage, node E hears both transmissions and produces the combination $d_1 \oplus d_2$ (bitwise XOR), which can be forwarded to the destination on the third path. Therefore, the destination receives three equations in two unknowns, where any two equations are solvable and are enough to recover the original data units. That is, two data units are proactively protected against a single node or link failure using only three paths, i.e., using 25% fewer paths than duplication. Note that the saving increases as the number of available paths increases.

1.5 Book Organization

This brief deals with techniques and strategies for applying network coding to achieve resilient operation of WSNs while achieving the two desired objectives, namely, agility in error recovery, and minimality of resources. This brief is organized into a number of chapters. Chapter 2 introduces a centralized approach for implementing network coding-based resilient operation of WSNs. The chapter introduces the theory behind this resilient operation in addition to practical implementation strategies. The strategies include protection against a single failure, and are then generalized to protect against multiple failures. Fast, but near optimal coding approaches are also described. A distributed approach for network coding-based resilient WSNs operation is the topic of Chap. 3. This approach tolerates both node and link failures, as well as packet losses. Coding strategies are also introduced for this approach. Chapters 2 and 3 are based on the assumption of using digital network coding, in which data packets are combined in the electronic domain. However, network coding can also be implemented by combining data in the electromagnetic domain. Two approaches are available for doing this, namely, physical-layer network coding, and analog network coding. Chap. 4 considers scheduling transmissions for achieving network coding protection using both digital and analog network coding under the centralized approach of Chap. 2. Scheduling strategies using digital and analog network coding are introduced. The performance of using digital network coding and analog network coding will be compared, which shows the merits of using analog network coding. Finally, Chapter 5 will conclude this brief.

References

1. I. F. Akyildiz and M. C. Vuran, "Wireless Sensor Networks", John Wiley and Sons, 2010.
2. Shuang-Hua Yang, "Wireless Sensor Networks Principles, Design and Applications", Springer, 2014.
3. O. M. Al-Kofahi and A. E. Kamal, "Survivability Strategies in Multihop Wireless Networks", IEEE Wireless Communications, Vol. 17, No. 5, Oct. 2010, pp. 71–80.
4. M. Younis, I. Sentruk, K. Akkaya, S. Lee and F. Senel, "Topology management techniques for tolerating node failures in wireless sensor networks: a survey", Elsevier Computer Networks, 58 (2014) 254–283.
5. R. Ahlswede, N. Cai, S. R. Li, and R. Yeung, "Network Information Flow", IEEE Transactions on Information Theory, Volume 46, No. 4, July 2000, pp. 1204–1216.
6. S. R. Li, R. W. Yeung, and N. Cai, "Linear Network Coding", IEEE Transactions on Information Theory, Vol. 49, No. 2, February 2003.
7. C. Fragouli and E. Soljanin, "Information Flow Decomposition for Network Coding", IEEE Transactions on Information Theory, Vol. 52, No. 3, March 2006.
8. S. Jaggi, P. Sanders, P. A. Chou, M. Effros, S. Egner, K. Jain, and L. Tolhuizen, "Polynomial Time Algorithms for Multicast Network Code Construction", IEEE Transactions on Information Theory, Vol. 51, No. 6, June 2005.
9. T. Matsuda, T. Noguchi and T. Takine, "Survey of Network Coding and Its Applications", IEICE Transactions on Communications, Vol. E94-B, No. 3, 2011, pp. 698–717.

Chapter 2
Network Coding-Based Resilient WSNs: The Centralized Approach

Link failures in Wireless Sensor Networks (WSNs) may occur due to severe channel fading, high levels of interference caused by other devices using the ISM band, or even physical damage to the network nodes or their antennas caused by the harsh surrounding environments. These problems may last for a considerable amount of time and they cannot be relieved using channel coding techniques or retransmissions. This chapter describes how to efficiently provide protection against such failures in a proactive manner while minimizing the used resources.

A centralized network coding-based scheme to provide proactive protection in a WSN will be introduced. This is done under the assumption that global network information is available, which may be the case in small size WSNs. With this assumption, this chapter makes it possible to focus on the theoretical aspects of the problem.

2.1 Problem Description

Let us consider the following motivating example shown in Figs. 2.1, 2.2, 2.3 and 2.4. In this network there are two sources, S_1 and S_2 that need to send two data units, b_1 and b_2, respectively to a sink node T. To provide proactive protection against a single link failure each source must use the network in a different time slot to have two edge-disjoint paths to the sink as shown in Fig. 2.2. This is because the minimum-cut between the sources and the sink is 3. However, unlike proactive protection, if reactive protection is to be provided, the two sources can use the network in the same time slot as shown in Fig. 2.3. If a failure takes place on one of the primary paths the affected source will have to detect the failure first, and then reroute its data to use the backup path through node A, which introduces delay and interrupts network operation. Now suppose that node A in the network shown in Fig. 2.1 is allowed to combine b_1 and b_2 (bitwise XOR), and send the

© The Author(s) 2015
O.M. Al-Kofahi, A.E. Kamal, *Resilient Wireless Sensor Networks*, SpringerBriefs
in Electrical and Computer Engineering, DOI 10.1007/978-3-319-23965-1_2

Fig. 2.1 Original graph

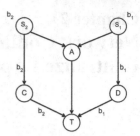

Fig. 2.2 1 + 1 protection

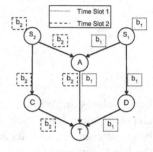

Fig. 2.3 1:2 protection

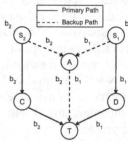

Fig. 2.4 Network coding

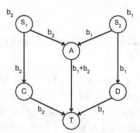

resulting symbol to the sink on the link (A, T), as illustrated in Fig. 2.4. This way, the two sources can use the network in the same time slot and still achieve proactive protection. If any of the three symbols sent to the sink is lost due to a link failure, the sink will still be able to recover the original data units. For example, assume that link (S_2, C) fails, the sink will receive $b_1 \oplus b_2$ on link (A, T) and b_1 on link (D, T), and it can recover b_2 by performing the bitwise XOR operation on the received symbols.

For the purpose of the development in this chapter, two types of nodes, namely, source nodes (the ones producing data), and coding nodes (the ones that may combine data) will be considered. From this perspective, the set of sources are referred to as U, and the set of coding nodes are referred to as L. To simplify the analysis, consider a WSN in which there is only one base station, and in which the nodes can be organized in t levels, where the minimum hop count between the base station and the nodes in level i is i (similar to [1, 2]). In general, it is better to combine the data units as early and as close as possible to the sources since this will reduce the used network resources. Therefore, we assume that U contains the nodes in some level j, and that L contains their direct neighbors in level $j - 1$. This assumption greatly simplifies the analysis, but does not compromise the theoretical results in any way as will be seen later.

In this chapter, the following questions are answered:

- How can network coding be used to provide protection against link (or path) failures in such many-to-one flow networks, while using the minimum possible number of paths?
- What are the necessary and sufficient conditions for such a solution to exist?
- If such a solution exists, how does it affect the network performance?

2.2 Network Coding-Based Resilience Approach

This section starts by developing the solution on a restricted network topology, which assumes the satisfaction of some connectivity and topology requirements (as will be stated below). Then, it is shown how to relax each of these requirements, and provide an appropriate generalization in Sect. 2.3.

2.2.1 Assumptions, Definitions and Notation

Since we are interested in the many-to-one flow from the sources in U to the base station, we can adopt the directed graph model in which a graph $G = (V, E)$ is used to represent the network. The set of vertices V represents the network nodes, and the set of edges E represents the available wireless links between network nodes, such that the edges are always directed from levels with higher indices to levels with lower indices. Considering the set of sources U and the set of coding nodes L, the edges will be directed from U to L since U contains the nodes in the next higher level to L. Taking that into account, we use the following assumptions and definitions:

1. Let $|U| = n$, and $|L| \geq n + 1$. In practice, this assumption is not always true. The reason to make such an assumption will become clear shortly. This assumption is relaxed in Sect. 2.3.4.

2. Let T represent the base station. Also, in the remainder of this chapter base station and sink may be used interchangeably since it is assumed that all nodes have the base station as their only destination.
3. All the links in the original graph G are of unit capacity unless stated otherwise, and there are no parallel links.
4. The minimum link cut between the nodes in L and the base station T is equal to $|L|$. Networks that do not have this property are discussed in Sect. 2.3.
5. The sub-graph induced by the nodes in U and L is bipartite. The general cases will be considered later in this chapter. This assumption can be naturally achieved in hierarchical or gradient-based routing protocols.
6. Only one link fails at a time.
7. All data packets have the same length.
8. G^T is the graph formed by: the nodes in U and L and all the links between them, a hypothetical sink node T' and hypothetical links from all the nodes in L to T'. G^T for the graph in Fig. 2.5 is shown in Fig. 2.6.
9. G^{ST} is the graph formed by: the nodes in U and L, and all the links between them, with a capacity of n assigned to each of these links, a hypothetical sink node T', hypothetical links with capacity of n from all nodes in L to T', a hypothetical source node S' and hypothetical links with capacity of n + 1 from S' to the nodes in U. G^{ST} for the graph in Fig. 2.5 is shown in Fig. 2.7.

Fig. 2.5 Graph G

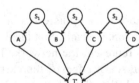

Fig. 2.6 Graph G^T

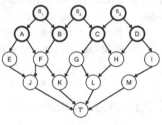

Fig. 2.7 Graph G^{ST}

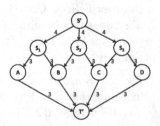

2.2.2 Sufficient and Necessary Conditions

To make the information flow from the n source nodes in U tolerant to a single link failure, we need to send $n + 1$ linear combinations (or equations) to the sink on $n + 1$ edge-disjoint paths. Moreover, to guarantee the successful recovery of the data, any n combinations from the $n + 1$ must be linearly independent (solvable). This way, the sink can recover the original n data units by solving any n from the $n + 1$ linear combinations, and since the $n + 1$ paths are edge-disjoint, a single-link failure will affect at most one path and only one combination will be lost. Note that similar to $1 + 1$ protection, there is no need to detect the failure. Note also, that similar to $1 : N$ protection only $n + 1$ paths are used.

The problem is divided into two sub-problems. The first deals with the needed information content in the linear combinations, i.e., how should the data units be incorporated in the combinations to guarantee the successful recovery of the original n data units in the case of a failure. The second is the coding problem to guarantee the linear independence of any n from the $n + 1$ linear combinations. This section focuses on the former and leaves the latter to Sect. 2.4. Thus, it is always assumed that the created combinations are linearly independent in this section.

Since we want to tolerate a single failure let us assume that $|L| = n + 1$ for now. From the assumptions above, each of the nodes in L has its own path to the sink that is edge-disjoint from the paths used by other nodes. Therefore, for simplicity, the original graph G can be replaced with G^T, where a path from a node in L to the sink is represented by a direct link. Taking this into account, the condition that will enable the L nodes to construct the $n + 1$ combinations that can tolerate a single link failure is that *any k nodes in U, must be connected to at least $k + 1$ nodes in L, for all values of k, where $1 \leq k \leq n$.* Consider the network in Fig. 2.8. The condition is satisfied for k = 1 since each node has two neighbors in L (or equivalently, has two paths to the sink). Also, the condition is satisfied for k = 3. However, it is not satisfied for k = 2 since S_1 and S_2 are only connected to two nodes in L. Therefore, if either of the links AT or BT fails, the sink will not be able to recover all three data units, because the other link that did not fail will be carrying the only combination of the two data units b_1 and b_2, while the sink needs at least two.

Fig. 2.8 Condition not satisfied for k = 2. S_1 and S_2 are only connected to two nodes in L

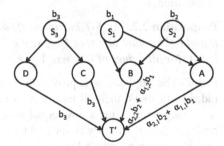

We now continue with proving that this condition is necessary and sufficient for the nodes in L to be able to construct the $n + 1$ combinations that can tolerate a single link (or path) failure.

Proposition 2.1. *The sink will recover the n data units even if one of the $n + 1$ combinations is lost, if and only if, any subset of nodes in U of size k is connected to a subset in L of size at least $k + 1$, for all values of k, where $1 \le k \le n$.*

Proof. In the previous scenario we viewed the data units from sources as variables, and the $n + 1$ nodes in L as combinations (or equations), and a variable is present in an equation if the corresponding source is connected to the node representing that equation.

The implication is proved by contradiction. Assume that the sink is able to recover the n data units, even if one of the $n + 1$ combinations is lost. But let there be a subset of U nodes of size k, that is connected to a subset of L nodes of the same size k. Then, the sink cannot randomly choose n combinations from the $n + 1$, because it **MUST** pick all the k combinations that were formed by the subset of L nodes mentioned above; otherwise, the k variables from the corresponding k nodes in U will only be present in $k - 1$ equations, i.e., they cannot be recovered. This contradicts the assumption that the sink is able to recover the original n data units if **ANY** of the combinations was lost, which concludes the proof of the implication.

To prove the converse, contradiction will also be used. Assume that any subset of nodes in U of size k is connected to another subset of nodes in L that is of size at least $k + 1$. But, there is a mandatory combination, which cannot be lost for the sink to be able to recover the original n data units. A combination is essential and can not be lost, if it leaves a set of equations of size say l with $l + 1$ unknowns, which are impossible to solve without that combination. But for this case to happen, there must have been some $l + 1$ nodes in U that are only connected to $l + 1$ nodes in L, which contradicts the original assumption, of having any k nodes in U connected to at least $k + 1$ nodes in L, for all values of k, where $1 \le k \le n$. ☐

Since all groups of nodes of all the sizes from 1 to n need to be checked, it may seem that checking the satisfiability of this condition is of order $O(2^n)$. It will now be shown how to check the condition above in polynomial-time with respect to the number of sources using a max-flow algorithm. The graph G^{ST} is used in the next proposition.

Proposition 2.2. *An S-T maximum-flow of at least $n(n + 1)$ is achievable in G^{ST}, if and only if, any subset of U of size k is connected to a subset in L of size at least $k + 1$, for all values of k, where $1 \le k \le n$.*

Proof. The implication is proved by contradiction. Assume the max-flow value is indeed $n(n + 1)$, then all the links from S to the n original sources are saturated (i.e., each one carries a flow equal to $n + 1$). And assume that there are some k nodes in U that are **only** connected to some k other nodes in L. The incoming flow to this component equals $k(n + 1) = kn + k$ while the outgoing capacity equals kn, which means that there are k units of flow that will be blocked from the sink, that is the max-flow $= n(n + 1) - k$ which contradicts the original assumption of the max-flow.

The converse is proved by contradiction. Suppose that any k nodes in U are connected to at least $k + 1$ nodes in L, but the maximum achievable flow was less than $n(n + 1)$, then there are some of the links from S to the nodes in U that could not be saturated. Assume only one of those links carried n units of flow to a node in U say node u. Then node u either has a single outgoing link, or is one of k nodes in U, that are connected to another set of k nodes in L (otherwise, it would have been able to forward this remaining unit of flow to the sink through an augmenting path on the residual network [3]). In both cases node u will violate the connectivity assumptions. Therefore, the max-flow must be $n(n + 1)$. This concludes the proof.

$\square$

2.3 Generalizations and Practical Considerations

In this section we generalize the conditions in order to cover any network topology, and to tolerate multiple failures. Then we consider the cases when $|L| > n + 1$ and when the network has a limited min-cut.

2.3.1 General Network Topology

So far, we have assumed that the graph induced by the nodes in U and L is bipartite. This assumption may not always apply. Therefore, we now generalize the above conditions for general network topology.

By carefully inspecting the condition of Proposition 2.1, one can see that the essence of the solution lies in the number of edge-disjoint paths from a group of sources to the sink. In the special case considered previously, each node in L represented one such path. Hence, Proposition 2.1 can be generalized in the following Theorem.

Theorem 2.1. *The sink will be able to recover the n data units even if ANY one from the n + 1 combinations is lost, if and only if, any subset of nodes in U of size k is connected to the sink through a set of edge-disjoint paths of size at least k + 1, for all values of k, where $1 \leq k \leq n$.*

Proof. The proof follows directly the same reasoning used in proving Proposition 2.1. $\square$

As an example to illustrate Theorem 2.1, consider a less restricted network topology, where links between sources are allowed, as shown in Fig. 2.9. It can be easily verified that the network in Fig. 2.9 satisfies the condition in Theorem 2.1. Moreover, this condition can be checked using the same idea in Proposition 2.2.

Fig. 2.9 Non-bipartite topology

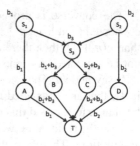

2.3.2 Multiple Failures

The necessary and sufficient conditions for the case of multiple failures can be derived from the above discussion, and are summarized in the following theorem:

Theorem 2.2. *The sink will be able to recover the n data units even if e link failures occur (i.e., at most e combinations are lost), if and only if, any subset of U of size k is connected to the sink through a set of edge-disjoint paths of size at least $k + e$, for all values of k, where $1 \le k \le n$.*

Proof. The proof follows directly the same reasoning used in proving Proposition 2.1. □

Although two generalizations have been discussed in the rest of the chapter the analysis of the baseline case that satisfies the assumptions in Sect. 2.2.1 will continue.

2.3.3 The Case of $|L| > n + 1$

Until now, it has been assumed that the number of nodes in L is exactly $n + 1$. The number of L nodes could be larger than $n + 1$. This however, does not invalidate our conditions and the above requirements will still apply. However, there are two issues that should be noted. First, since $|L| > n + 1$ the number of produced combinations may be more than $n + 1$ depending on the connectivity between U and L. One extreme case is when each node in L is connected to only one node in U, which will cause the number of combinations to be $2n$. The second issue is that when the number of combinations is larger than $n + 1$, we cannot randomly select any n combinations to recover the original data units. The network shown in Fig. 2.10 gives a good example, where the minimum number of combinations is $n + 2$. In this network, selecting four combinations randomly may not cover all the data units. For example, if the combinations created in nodes C, D, E and F were chosen by the sink to calculate the original four data units, b_1 cannot be recovered. However, if any of the above mentioned four combinations was replaced by either

Fig. 2.10 Minimum
combinations = n + 2

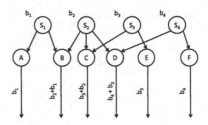

of the combinations from A or B, the sink will be able to recover all the original
data units.

2.3.4 Networks with Limited Minimum Cuts

So far, it has been assumed that the number of edge-disjoint paths from L to T is
larger than or equal to $n + 1$. Obviously, this may not be the case always. From
Menger's theorem [4], the maximum number of edge-disjoint paths between the
nodes in L and the sink is equal to the minimum edge cut (min-cut for short) between
L and T. Let h denote the min-cut. If $h \geq n + 1$, then the developed approach can
be applied directly, and the combinations formed in L can be forwarded to the sink.
On the other hand, if h is less than $n + 1$, then the formed combinations cannot be
forwarded as is, and the above approach must be modified.

Let us assume that $h < n + 1$, then the sink cannot receive more than h
combinations. That is, to protect against a single failure, the combinations should
not contain data from more than $h - 1$ sources. Therefore, we divide the n sources
into groups of size $h - 1$ sources each, and then choose a set of feasible groups that
covers all the sources. It is assumed that the groups are time multiplexed, and a group
of sources is defined to be feasible if it satisfies the condition in Proposition 2.1.
A set of groups is defined to cover all the sources, if each source is present in at
least one of the groups in the chosen set.

The way to choose the covering set of feasible groups must take the following
into consideration:

1. The degree of disjointedness between groups, which affects the fairness and the
   rate at which the sources transmit, as will be seen in Sect. 2.5.
2. The used network resources.

As an illustration consider the network in Fig. 2.11, which contains four sources.
The maximum number of edge-disjoint paths (h) from L to T in this network is 3.
Therefor, the largest possible group of sources that can be protected together, will
be of size at most 2. In this example all the groups of size two are feasible according
to Proposition 2.1. Let us now compare the following three sets of groups:

Fig. 2.11 A network with $h = 3$

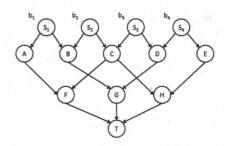

1. $Set_1=\{\{S_1S_2\}, \{S_1S_3\}, \{S_1S_4\}\}$: In this set, the rate of S_1 is 1 *symbol/time slot* since it is present in all groups. While the rate of other sources is $\frac{1}{3}$ *symbol/time slot*, which is unfair.
2. $Set_2=\{\{S_1S_3\}, \{S_2S_4\}\}$: this choice of groups achieves better fairness, where the bandwidth is equally divided and all the sources transmit at a rate of $\frac{1}{2}$ *symbol/time slot*.
3. $Set_3=\{\{S_1S_2\}, \{S_3S_4\}\}$: this is the best solution, because not only we achieve better fairness than Set_1, but also we use fewer network resources than Set_2 since each of the groups uses only three links to forward data to the minimum cut edges, compared to four links in the group of Set_2.

The problem of choosing the smallest set of feasible groups to cover all sources can be proved to be NP-complete through a reduction from the **K-Set Cover** problem [5].

2.3.5 MILP Formulation

In this section the problem of source grouping is formulated as a mixed integer linear program. First, we define the following:

1. $s(k)$ source number k, where $1 \leq k \leq n$.
2. M the maximum number of groups which equals $n - h + 2$.
3. f_{ij}^{kc} the flow of source k in group c on edge (i,j).
4. z_{ij}^c a binary variable which is equal to 1 if the edge (i,j) carried flow for group c and 0 otherwise.
5. g_c^k a binary variable which is equal to 1 only if source k was in group c and 0 otherwise.

To guarantee the feasibility of each group, we use the same method presented in Proposition 2.2. That is, we assume each source has a flow of h units, and that the capacity of all edges is $h - 1$. This way, a group will be feasible if a flow of $h(h-1)$ can be forwarded to T. Taking this into account, the linear integer program is as follows:

Objective function:

$$\text{Minimize} \sum_{c=1}^{M} \sum_{\forall (i,j) \in E} z_{ij}^c \tag{2.1}$$

Subject to:

$$\sum_{\forall j:(s(k),j) \in E} f_{s(k)j}^{kc} = g_c^k.h \ , \ \forall k, c \tag{2.2}$$

$$z_{ij}^c - \frac{\sum_{k=1}^{n} f_{ij}^{kc}}{h-1} \geq 0 \ , \ \forall c, (i,j) \in E. \tag{2.3}$$

$$\sum_{\forall i:(i,j) \in E} \sum_{k=1}^{n} f_{ij}^{kc} = \sum_{\forall i:(j,i) \in E} \sum_{k=1}^{n} f_{ji}^{kc} \tag{2.4}$$

$$\forall c, j \notin \{T, s(1), \ldots, s(n)\}$$

$$0 \leq \sum_{k=1}^{n} f_{ij}^{kc} \leq h-1 \ , \ \forall c, (i,j) \in E \tag{2.5}$$

$$\sum_{c=1}^{M} g_c^k = 1 \ , \ \forall k \tag{2.6}$$

$$\sum_{k=1}^{n} g_c^k \leq h-1 \ , \ \forall c \tag{2.7}$$

The objective in (2.1) is to minimize the number of used links for each group. Constraint (2.2) says that if source k was participating in group c the outgoing flow from it must be equal to h in the time slot for that group. Constraint (2.3) forces z_{ij}^c to be equal to 1 if the flow on edge (i,j) was not 0. Constraint (2.4) says that in a certain group (i.e. at a certain time slot) the amount of flow (of all sources) entering a node equals the amount of flow leaving that node. Constraint (2.5) says that the sum of flow of all sources in a certain group cannot exceed the capacity of any link which is equal to $h-1$. Constraint (2.6) ensures that each source participates in one group only, and (2.7) guarantees that a group contains no more than $h-1$ sources.

This MILP guarantees fair bandwidth sharing, i.e., a source cannot transmit again unless all other sources have transmitted. This is ensured by constraint (2.6) that forces each source to participate in one group only. As will be shown later in Sect. 2.5, a source might have the opportunity to transmit more than once without affecting the throughput of other sources; let us call this opportunistic transmission. The MILP can be modified for opportunistic transmissions as follows:

The objective function should be:

$$Minimize \sum_{c=1}^{M}(\sum_{k=1}^{n} g_c^k + \sum_{\forall (i,j) \in E} z_{ij}^c) \tag{2.8}$$

with the following modifications on constraints (2.6) and (2.7):

$$\sum_{c=1}^{M} g_c^k \geq 1 , \forall k \tag{2.9}$$

$$\sum_{k=1}^{n} g_c^k = h - 1 , \forall c \tag{2.10}$$

Now the bandwidth is utilized by constraint (2.10) that sets the size of all groups to its maximum size $h - 1$, and a source is allowed to participate in more than one group by constraint (2.9).

2.4 Coding

In the previous sections, linear independence between the linear combinations produced in L was assumed. In this section, it will be shown how to achieve this independence between combinations through using $\{0, 1\}$ coefficients. This reduces all operations to bit-wise XOR operations, and simplifies the coding and decoding processes.

Achieving independence using $\{0, 1\}$ coefficients depends solely on how one composes each combination from only the data units, i.e., a data unit is present in a combination if its coefficient is 1, and a data unit is not present if its coefficient is 0. For instance, in Fig. 2.4, the three combinations that were sent to the sink are, $C_1 = b_1$ (coefficient of b_2 is 0), $C_2 = b_2$ (coefficient of b_1 is 0) and $C_3 = b_1 + b_2$. In the following sections we assume that the connectivity condition of Proposition 2.1 is satisfied, and it will be shown how to decide on the data units composing each of the linear combinations, through finding simple paths and trees.

2.4.1 Path Coding and Tree Coding

Consider the bipartite graph induced by the nodes in U and L. Assume there exists a an undirected path that has both ends in L. Note that since the graph is bipartite, the path will alternate between the nodes in U and the nodes in L. Assume also, that the path includes all the nodes in U. For such path, it is clear that any k nodes from U

on that path have at least $k + 1$ neighboring nodes from L that are also on the same path. In addition, note that the number of L nodes on such a path is the minimum number of nodes that satisfies the connectivity conditions, because each source has only two neighboring nodes in L.

Such a path not only finds a set of nodes in L that satisfy the connectivity conditions, but also helps in the assignment of the coding coefficients to create the needed linearly independent combinations. To illustrate the benefits of coding according to the connectivity on a path, consider the example in Fig. 2.12, where there are 4 source nodes in U, and 5 nodes in L. If every node in L XORs all the data units it receives, we will have dependent combinations like $\{b_4 + b_3\}$ and $\{b_4 + b_3\}$ (or $\{b_1 + b_2\}$ and $\{b_1 + b_2\}$), thus invalidating the linear independence requirements.

However, consider the simple undirected path $\{A, S_1, B, S_2, C, S_3, D, S_4, E\}$ that is represented by the solid edges in Fig. 2.13. If a coding coefficient of 1 is assigned to the links on the path (solid), and a coding coefficient of 0 is assigned to the links not on the path (dashed), then linear independence will be guaranteed, since any two combinations cannot have more than one element (i.e., data unit) in common.

Of course, it may not be always possible to find such a simple path that covers all the nodes in U. However, a path is only but a tree with two leaf nodes, or a tree with a root and one leaf node (depends on where the root of the tree is). Therefore, path coding can be generalized to tree coding, where we need to find a tree that covers all the nodes in U, with all of its leaf nodes in L. Note that requiring the leaves to be in L is only to guarantee that each node in U has two neighbors in L. Assigning tree links a coefficient of 1, and non-tree links a coefficient of 0 guarantees linear independence as stated in the next theorem and as illustrated in Fig. 2.14 (Fig. 2.15 clarifies the underlying tree structure).

Fig. 2.12 Bad coding: linear independence of any n combinations is not satisfied

Fig. 2.13 Path coding: linear independence is satisfied in any n combinations

Fig. 2.14 Tree coding: linear independence is satisfied in any n linear combinations

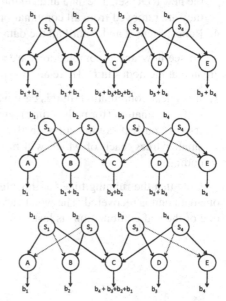

Fig. 2.15 Underlying tree

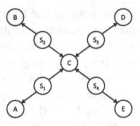

Theorem 2.3. *If the linear combinations at the nodes in L are created according to their connectivity with the source nodes in U on the coding tree, i.e., a link on the tree is assigned a coefficient of 1 and a link not on the tree is assigned a coefficient of 0, then any n combinations from the resulting n + 1 combinations are linearly independent.*

Proof. A direct proof is used to prove this implication. We prove that any n combinations from the $n + 1$ are linearly independent by proving that they are solvable by constructing an algorithm to solve for the n data units. In the algorithm the term *"leaf combination"* refers to a combination created at a leaf node in the tree, which will be a trivial combination that consists of a single data unit. The algorithm works as follows:

1. Put all the data units from the leaf combinations in a set; let us call it the *Recovered Data Units Set*, or the *RDU set* for short.
2. Remove all data units in the RDU set from the remaining combinations. This is done through XORing a data unit with all the combinations that it participates in.
3. After the previous step, a new set of data units will be recovered. These compose the new RDU set. The data units in the old RDU set will not be used further since they are removed from all combinations.
4. Repeat Steps 2 and 3 until all the data units are recovered.

To see how any n combinations are solvable, remove one of the combinations created at the nodes in L. There are two possibilities for this combination:

1. It is a leaf combination: in this case we are guaranteed to have at least one other leaf combination (for path coding), and the algorithm can start from it.
2. It is a non-leaf combination: in this case the coding tree is divided into two smaller trees, each of which will have at least one leaf combination (for path coding).

Note that the running time for this algorithm is $O(n)$, since in each step at least one data unit is recovered. The worst case occurs when the coding tree is a path and one of the leaf combinations is lost. □

Algorithm 2.1 Construct a Coding Tree

1: **while** there are leaves from U **do**
2:     Pick a leaf node from U in the tree, say u
3:     Find one of u's neighbors in L say x //other than u's parent
4:     Call **ModTree**(u, x)
5: **end while**

Algorithm 2.2 ModTree(u,x)

1: Connect u to x, this will create a cycle, say C
2: Traverse the nodes on C, until we reach a node in U, say node v, that has a neighbor w not on the cycle.
3: **if** w is already connected to v **then**
4:     Cut the cycle directly before or after v
5:     **return**
6: **else**
7:     Cut the cycle before or after v
8:     Call **ModTree**(v, w)
9: **end if**

2.4.2 Constructing a Coding Tree

A coding tree can be constructed using the following three steps:

1. Build a tree rooted at a node in L. This can be a depth-first search or a breadth-first search tree (DFS or BFS). This step is of order $O(|E|)$.
2. Modify the tree to guarantee that all leaf nodes are in L.
3. If $|L| > n + 1$, trim extra leaves. This operation is of order $O(|V|)$.

The involved step is step two since steps one and three can be easily accomplished. Algorithms 2.1 and 2.2 describe a procedure to do the modification in the second step.

In Algorithm 2.1, we search the tree for a leaf node that falls in U. Upon finding such a leaf node u, we look for a neighbor x in L for node u that is different from the parent of u. It is guaranteed to find such a neighbor x, because each single node in U is connected to at least 2 nodes in L (Proposition 2.1). After finding u and x, the procedure **ModTree** adds the link between them to the tree creating a cycle C. Then, it traverses the nodes on C to find a node v in U that has a neighbor w in L not on C. Again, we are guaranteed to find such a node v that has such a neighbor w, because any k nodes in U are connected to at least $k + 1$ nodes in L, and since the cycle is composed of equal numbers of nodes from both U and L, then there must be a U node on this cycle that has a neighbor in L not on the cycle. If v was connected to w on the tree, we cut the cycle before or after v, to make the graph a tree again. On the other hand, if v and w are not connected on the tree, we recursively call **ModTree** (shown in Algorithm 2.2). As an illustration consider the network in Fig. 2.16, the resulting DFS-Tree (Depth First Tree) rooted at node C is shown in Fig. 2.17. The nodes that will be found when running Algorithm 2.1 and two

Fig. 2.16 Network with
$n = 3$

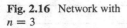

Fig. 2.17 DFS-Tree

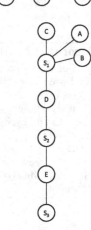

Fig. 2.18 Two iterations

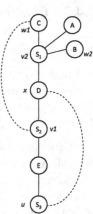

iterations of **ModTree** are shown in Fig. 2.18, the cycles and the edges that are marked to be cut are shown in Fig. 2.19, and the final result after trimming the extra leaf nodes is shown in Fig. 2.20.

There can be at most $n - 1$ leaf nodes in U, and the recursive call to **ModTree** can be done at most n times. Hence, the running time of Algorithm 2.1 is of order $O(n^2)$.

Fig. 2.19 Making the
modification

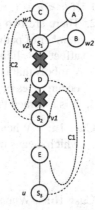

Fig. 2.20 Result

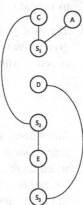

2.5 Network Performance

Sending redundant data reduces the effective data rate. This section studies the
effect of network coding-based protection on the effective data rate in the following
cases:

- **Case 1:** Fair bandwidth sharing, with no protection.
- **Case 2:** Fair bandwidth sharing, with protection.
- **Case 3:** Opportunistic transmission, with no protection.
- **Case 4:** Opportunistic transmission, with protection.

Fair bandwidth sharing means that a source does not transmit again until all
other sources have transmitted, and opportunistic transmission means that a source
transmits whenever it has an opportunity to do so. In the discussion, the rate $\mathscr{R}$ is
defined as the number of data units that can be received by the sink per unit time.
Also the following is assumed:

1. If $h < n + 1$, sources will be divided into groups, as in Sect. 2.3.4.
2. There are two edge-disjoint paths from every source to the sink, i.e., minimum number of paths is used. This assumption will simplify the analysis, and will not affect its validity. This is because the rate is affected by the number of unique data symbols that can be recovered from the combinations, but not by the number of occurrences of the data symbols in the combinations.
3. It is assumed that the selected groups are feasible as defined in Sect. 2.3.4.

Case 1a: When $h \geq n$, the sink can receive all the n data units at the same time, which means that:

$$\mathscr{R} = n$$

Case 1b: When $h < n$, the sources should be grouped, which will cause the rate to vary at the sink depending on the way the sources were divided:

1. The best grouping scenario is when the sources are sorted in disjoint groups, giving rise to $\lceil \frac{n}{h} \rceil$ groups, i.e., $\lceil \frac{n}{h} \rceil$ time units are needed for all the sources to be covered, which means that:

$$\mathscr{R} = \frac{n}{\lceil \frac{n}{h} \rceil} \leq \frac{n}{\frac{n}{h}} = h$$

2. Since every source has two-edge disjoint paths to the sink, then any source can reach two different edges in the min-cut. The worst grouping scenario occurs when there are $h - 2$ min-cut edges that can be only reached by $h - 2$ sources, forcing the remaining $n - h + 2$ sources to use just the remaining two min-cut links. Because we assume no protection in this case, each source can use one edge in the min-cut, hence, the $n - h + 2$ sources can be divided into $\lceil \frac{n-h+2}{2} \rceil$ pairs, each pair when combined with the other $h - 2$ sources will form a group. Which produces $\lceil \frac{n-h+2}{2} \rceil$ similar groups that only differ in two elements. The network in Fig. 2.21 shows an example, where the sources $S_1, S_2, \ldots, S_{h-2}$ are present in all the selected groups (although we assume that they transmit in only one time slot out of the $\lceil \frac{n-h+2}{2} \rceil$), and the sources from S_{h-1} to S_n can connect to the sink only through the last two min-cut links. In this case the $n - h + 2$ sources will share these two links in $\lceil \frac{n-h+2}{2} \rceil$ time slots. Which means that:

$$\mathscr{R} = \frac{n}{\lceil \frac{n-h+2}{2} \rceil} \leq \frac{n}{\frac{n-h+2}{2}} = \frac{2n}{n - h + 2}$$

It can be seen that the last two cases are equivalent when $h = 2$, and they both give $\mathscr{R} = 2$.

Case 2a: When $h \geq n + 1$ the sink can receive the $n + 1$ combinations at the same time and recover all the n data units using any n combinations, that is:

$$\mathscr{R} = n$$

Fig. 2.21 Worst case grouping

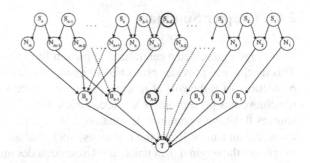

Case 2b: When $h < n + 1$ the sources should be divided into groups, and the rate at the sink will depend on how the grouping was accomplished:

1. As before, the best grouping is when the sources can be divided into disjoint groups, but in this case, since protection is assumed to be provided, the groups will be of size $h - 1$, thus producing at most $\lceil \frac{n}{h-1} \rceil$ groups. Hence, the rate will be:

$$\mathscr{R} = \frac{n}{\lceil \frac{n}{h-1} \rceil} \leq \frac{n}{\frac{n}{h-1}} = h - 1$$

2. The worst case grouping is similar to that discussed in Case 1b, but in this case since protection is assumed, the $n - h + 2$ sources mentioned earlier will not be divided into pairs, rather, each of which will be active alone with the other $h - 2$ sources, thus giving rise to $n - h + 2$ groups, each of size $h - 1$. The network in Fig. 2.21 still gives a valid example. The rate in this case is:

$$\mathscr{R} = \frac{n}{n - h + 2}$$

Note that the last two cases are also equivalent when $h = 2$, and give $\mathscr{R} = 1$, since each source must use two edge-disjoint paths to forward data to the sink.

Case 3: The calculations from Case 1 are still valid for this case, except when $h < n$ with worst case grouping, in which the rate at the sink will equal h, since the $h - 2$ sources are allowed to transmit in every one of the $\lceil \frac{n-h+2}{2} \rceil$ time slots. To capture the difference, we should consider the rate at the sources, where it was $\frac{1}{\lceil \frac{n-h+2}{2} \rceil}$ for each of the n sources in Case 1. However, in this case, $\mathscr{R} = 1$ for each of the $h - 2$ sources that are repeated in all groups and $\frac{1}{\lceil \frac{n-h+2}{2} \rceil}$ for each of the $n - h + 2$ sources that can only use two min-cut links.

Case 4: The only difference between this case and case 2, is when $h < n + 1$ with worst case grouping, in which the rate at the sink will always be equal to $h - 1$. Again, to see the difference we should consider the rate at the sources, where it will be 1 for each of the $h - 2$ common sources, and $\frac{1}{n-h+2}$ for each of the $n - h + 2$ sources that share only two min-cut edges.

2.6   Chapter Summary

This chapter presented a centralized approach for network coding-based protection. This mechanism provides protection to many-to-one flows at the speed of proactive protection but at the cost of reactive protection. Necessary and sufficient conditions to achieve this protection for n source nodes were derived. Basically, for a set of n sources that have a common destination, it was proven that to be able to send $n + 1$ linear combinations (from the n sources) such that any n combinations are enough to recover the original data units, any k source nodes must have at least $k + 1$ edge-disjoint paths to the common destination for all values of k, where $1 \leq k \leq n$. Note that sending $n + 1$ combinations on $n + 1$ disjoint paths guarantees protection against one failure. The study was generalized, and conditions to cover multiple failures were derived, and it was shown how to deal with networks with limited minimum cuts. In addition, a polynomial-time algorithm to perform binary network coding (i.e., using $\{0,1\}$ coefficients) was introduced. This simplifies the decoding process where only XOR operations are needed to recover the original data units. Finally, the effect of the network coding-based protection approach on the performance of the network was studied, and the data rate with and without protection was calculated.

References

1. C. Schurgers, S. Mani, "Energy efficient routing in wireless sensor networks.", In Military Communications Conference, 2001. MILCOM 2001. Communications for Network-Centric Operations: Creating the Information Force. IEEE, vol. 1, pp. 357–361. IEEE, 2001.
2. P. Quang and D. Kim, "Enhancing real-time delivery of gradient routing for industrial wireless sensor networks", Industrial Informatics, IEEE Transactions on 8, no. 1 (2012): 61–68.
3. T. Cormen, C. Leiserson, R. Rivest and C. Stein. *Introduction to Algorithms.* The MIT Press; 3rd edition (July 31, 2009).
4. J. Kleinberg and E. Tardos. *Algorithm Design.* Addison Wesley, 2005.
5. M. Garey and D. Johnson. *Computers and Intractability: A Guide to the Theory of NP-Completeness.* WH Freeman New York, 1979.

Chapter 3
Network Coding-Based Resilient WSNs: The Distributed Approach

This chapter introduces a mechanism to suite the needs of large size WSNs. The technique uses deterministic binary network coding in a distributed manner to enhance the resilience of Sensor-to-Base information flow against failures, including packet losses. The approach is a distributed version of the coding tree algorithm presented in Chap. 2. This version operates using local node information only, and does not make any assumptions on the source nodes that will encode their data units together. The concept of relative indexing is introduced, which facilitates the construction of coding vectors for this distributed code. A simple routing protocol is introduced to enable the coding mechanism to tolerate node and link failures.

3.1 Tolerating a Single Loss

This section defines the rules of operation that allows a group of sensor nodes to collaboratively protect their data units against a single loss. A node will be able to operate using its local information only, which makes the proposed mechanism fully distributed. Before discussing the details of node operation, the assumptions and used notation are clarified.

3.1.1 Assumptions and Notation

As will be seen later, the proposed mechanism is quite general and can work in almost any WSN. Nevertheless, some assumptions are made to help in streamlining the discussion, such as:

© The Author(s) 2015
O.M. Al-Kofahi, A.E. Kamal, *Resilient Wireless Sensor Networks*, SpringerBriefs
in Electrical and Computer Engineering, DOI 10.1007/978-3-319-23965-1_3

1. The sensor nodes in the wireless sensor network are organized into levels or rings around the base station (BS), such that, the minimum hop count between the sensor nodes in ring i and the BS is i hops. An example is shown in Fig. 3.1, where nodes d, e and f are in the first ring, nodes c and b are in the second ring and finally node a is in the third ring around the BS. The terms ring and level are used interchangeably throughout the chapter.
2. Routing a packet ensures its progress towards the sink in each transmission, i.e. a packet gets closer to the sink by one hop after each transmission.

In general, a WSN can work in different modes, including periodic, event-based, and query-based. Although the proposed mechanism works in any mode, it performs better in periodic sensing mode (where sensors generate and send data periodically) since it may increase the chances of combining data units. Therefore, periodic sensing is assumed. Finally, the data units from all sensor nodes are assumed to be equal in size, i.e., they have the same length. This assumption can be easily relaxed since if data units had different lengths, shorter ones can be padded. As for the notation, the following are used:

1. l_u denotes the level of node u.
2. d_u denotes the data unit for sensor node u.
3. Throughout the chapter p denotes a certain packet, and d_p is reserved to only represent the information symbol carried in packet p.

3.1.2   Rules of Operation

To protect against a single packet loss, a node u can send two copies of its data unit d_u to the BS, so that if a packet is lost the other would arrive. Duplicating data units in this manner will waste 50 % of the network resources (bandwidth and energy) to provide redundant data. However, note that, in a WSN, these copies are forwarded using multi-hop communication, which opens up opportunities to combine them with other data units along the route to the BS. In general, at any instance of time t_i, the nodes in a wireless sensor network can be divided into the following three types:

1. **Type 1.** A source with no data to relay. Such a node has generated its own data (a source), but does not have any data from other nodes to relay.
2. **Type 2.** A source with data to relay. Such a node has generated its own data, and has some data from other nodes that needs to be relayed too.
3. **Type 3.** Just a relay with no data of its own.

Note that the type of a node changes in time according to its status, i.e., if node u is of Type 3 at time t_i, it may become a Type 1 or Type 2 node at t_{i+1} if a data unit is generated locally. To tolerate a single loss using network coding in a distributed manner, each node operates according to its type as follows:

Fig. 3.1 Protecting data from
two sources (a and b) against
a single packet loss

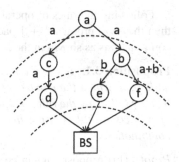

1. **Type 1.** Assume node u only has its own data unit d_u, and has no packets from
   other sensors to relay. Then node u just sends two copies of d_u to be able to
   tolerate a single loss, e.g, node a in Fig. 3.1.
2. **Type 2.** Assume node u has its own data unit d_u, and in addition has received
   another packet, say p, which needs to be relayed (e.g., node b in Fig. 3.1). Then
   node u produces the following two packets:

   (a) **Packet 1.** Contains the bitwise XOR of d_u and d_p.
   (b) **Packet 2.** Contains only d_u.

3. **Type 3.** Assume node u has no data of its own, and has received a packet p that
   it needs to relay. Then it just forwards p as is, e.g., nodes c, d, e and f in Fig. 3.1.

In Fig. 3.1, node a is the only node of Type 1, which sends two copies of its data
unit to the BS. As these copies are forwarded, one of them passes through node b
which happens to have generated a data unit of its own that is yet to be forwarded
to the BS. In this scenario, node b is of Type 2, therefore, it sends two packets as
described in the rules of operation. One packet contains the data unit of b, and the
other combines (XORs) d_b with d_a. The second copy of a's data unit passes through
c which have not generated any data yet (a Type 3 node), and thus, forwards d_a as is.
All other nodes are Type 3 nodes and relay the received packets as is.

Note that in this example there are two sources (a and b), and that there is a total
of three packets generated (a, b, and a+b). In general, if Type 1 node u sends two
copies of d_u to the sink. Only Type 2 nodes can collaborate with u and combine
there data units with d_u. According to the rules of operation, a Type 2 node can
combine its own data unit with other packets that can be either native packets (from
nodes of Type 1 or 2) or coded packets (from Type 2 nodes). Suppose that as the
two copies of d_u are forwarded, $k - 1$ of the forwarding nodes had data units of their
own (i.e., Type 2 nodes) that also need to be sent to the BS. Following the rules of
operation, a Type 2 node will increases the total number of packets by only one.
Therefore, $k + 1$ packets will be generated, two copies of d_u plus the $k - 1$ other
data units. Let us call this operation a *forwarding process initiated by node u*, or for
short, an *F-process* initiated by a Type 1 node, u.

Following the rules of operation, if k sources are involved in a certain F-process then they will produce $k + 1$ packets such that any k of them carry a solvable set of combinations as shown in the following proposition:

Proposition 3.1. *If s is the kth source to participate in an F-process initiated by node u, in which the number of involved sources until now is $k - 1$ (including u), then by following the rules of operation described above, s will increase the number of packets to $k + 1$, such that any k packets from them carry a solvable set of combinations.*

Proof. This proposition can be proven by induction. The proof needs to show that if any $k - 1$ from the current k combinations (produced by the $k - 1$ sources) are solvable, then the participation of the kth source will produce $k + 1$ combinations such that any k of them are solvable.

The *basis step* is when the F-process is first initiated at node u. By following the rules of operation, node u will send two copies of d_u. Thus, there is one source ($k = 1$) and two combinations ($k + 1 = 2$), which are trivial combinations in this case, such that any one (k) of them is solvable. This is obvious since each combination has only one data unit, so if one packet is lost the other is sufficient to recover d_u.

To prove the *inductive step*, assume that $k - 1$ sources have transmitted and any $k - 1$ from the current k combinations are solvable. Node s can participate only if it is of Type 2, i.e., it received one packet (say p) from the current k packets and it has its own data unit d_s that must be sent to the BS. By following the rules, s will produce two packets: (1) a packet containing the XOR of d_s with d_p (note that the number of total packets is still k since p and d_u are merged into one packet), and (2) a packet containing d_s [i.e., the number of packets is increased by (1)]. To complete the proof, it needs to be shown that losing either one of the newly created packets will leave k solvable combinations. If the first packet is lost, then the second packet will be sufficient to recover d_s, and from the starting assumptions the remaining $k-1$ combinations will be sufficient to recover the remaining $k - 1$ data units. Now if the second packet is lost, then from the assumptions all the data units in d_p (and thus d_p itself) can be recovered using the $k - 1$ combinations other than $d_s \oplus d_p$, and then d_s can be recovered by $d_p \oplus (d_s \oplus d_p)$. Finally, if a packet other than those produced by s is lost, d_s can be recovered from the second packet produced by s, which leaves $k - 1$ solvable combinations in $k - 1$ unknowns.   □

A node of Type 1, will initiate a process that will involve a large number of nodes of the other two types. It is important to distinguish the packets belonging to the same F-process initiated by a node of Type 1. This is because only the combinations of the same F-process can be decoded together. To identify the packets belonging to a certain F-process, a new field is added in all the packets generated by this process. Let this field be called the *"initiator ID"* or *IID* for short. A node of Type 1 will put its ID and a timestamp in this field, the timestamp is needed to distinguish between the F-processes from the same initiator. A node of Type 2 (say u) that XORs d_p with its data unit will copy the IID from p and put it in both generated packets.

Furthermore, if a Type 2 node receives two or more packets with the same IID it must not combine its data with more than one of them, since this may produce dependent (unsolvable) combinations.

Let P_i^j be the set of packets leaving level i (i.e., generated by level i, or just forwarded by it) that belong to the process initiated by node j, then $|P_i^j|$ is equal to:

$$|P_i^j| = 1 + \sum_{k=i}^{l_j} S_k^j = 1 + N_i^{l_j} \tag{3.1}$$

where S_k^j is the number of source nodes (Type 2) in level k that are involved in the process initiated by j, and $N_i^{l_j}$ is the number of source nodes in the levels from i to l_j that are involved in the process initiated by node j. Of course, there is only one node of Type 1 in the process initiated by node j, and that is j itself. Note that this is a huge reduction in the total number of packets compared to pure duplication-based approaches, where the number of packets will be $2N_i^{l_j}$.

There should be a limit on the number of combinations that can be produced by an F-process. Let h denote this limit. For now, h is chosen to be the minimum node cut in the network under consideration, and more on the selection of h will be discussed later in Sect. 3.5. In a sufficiently dense and uniformly distributed WSN, the minimum node cut is usually the number of one hop neighbors of the BS. Therefore, no more than $h - 1$ sources can be involved in any process. To insure this, a new field is added to the packet format. Let this field be called the *"Number of Remaining Combinations"* or *"NRC"* for short. Taking IID and NRC into account, the operation of Type 1 and Type 2 nodes is redefined as follows (the operation of Type 3 nodes does not change):

1. **Type 1.** Assume node u only has its own data unit d_u, and has no data units from other sensors to relay. Then node u just sends two copies of d_u, such that in both packets the values of $IID = u$ and $NRC = \lfloor h/2 \rfloor$.
2. **Type 2.** Assume node u has its own data unit d_u, and in addition has received another packet, say p, that it needs to relay. The operation of u depends on the NRC value of p as follows:

   (a) If $NRC(p) \geq 2$ node u produces the following two packets, such that, in both packets $IID = IID(p)$ and $NRC = \lfloor NRC(p)/2 \rfloor$.

      (i) **Packet 1.** Contains $d_u \oplus d_p$.
      (ii) **Packet 2.** Contains only d_u.

   (b) However, if $NRC(p) = 1$, node u acts as a node of Type 3, i.e., just a relay.

Following the new rules of operation, an F-process will have at most $h - 1$ sources and will produce at most h packets. It is worth mentioning that by taking the floor when updating the NRC value, h is made equivalent to the largest power of 2 that is less than or equal to h. For example, consider the network in Fig. 3.2 where $h = 7$. Node a is the initiator node (Type 1 node), by following the rules of operation a will

Fig. 3.2 NRC update for
h = 7

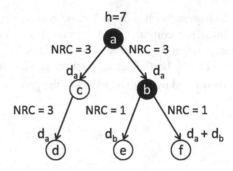

send two packets carrying d_a to b and c with $IID = a$ and $NRC = \lfloor \frac{7}{2} \rfloor = 3$. Node c is a Type 3 node, i.e., it will relay only and will not change the value of the IID or the NRC. Node b is a Type 2 and according to the rules of operation it will produce two packets with $IID = a$ and $NRC = \lfloor \frac{3}{2} \rfloor = 1$. It is obvious that starting with $h = 7$ we can have at most four packets produced, which is equivalent to starting with $h = 4$ (i.e., the closest power of 2 less than 7). Therefore, in the remainder of this chapter we will always assume that h is a power of 2.

Note that assigning the NRC value as described in the rules of operation is not optimal in terms of maximizing the number of participating sources in a certain F-process. For example, in Fig. 3.2, assume that similar to node c node d is also a Type 3 node and the packet forwarded from d will reach the BS on a path composed only of Type 3 nodes. However, assume that node f has a data unit of its own that it wants to combine with the packet received from b. Node f will not be able to participate since the NRC value in the received packet is 1. Therefore, if node a knew beforehand the source nodes that will relay its packets, it can set the NRC to 1 in the packet sent to c, and to 6 (which is equivalent to 4) in the packet sent to b. Nevertheless, since we assume that a node operates using only its local information, and the deployed network is dense and uniformly distributed, a realistic assumption that an initiator node can make is that both packets will pass through the same number of sources approximately, and thus divide h by 2. Using simulation, in Sect. 3.6 we verify that using this assumption results in a performance that is not much worse than the optimal.

3.2   Tolerating Multiple Losses

In this section, a simple extension to the mechanism introduced in Sect. 3.1 is proposed to provide protection against e losses. Basically, if a data unit needs to be protected against e losses, it must be incorporated in at least $e + 1$ combinations. According to the rules of operation, only nodes of Types 1 or 2 can generate combinations, where each time a node participates in an F-process it incorporates its data unit in two combinations. Therefore, if e is an odd number, a source node

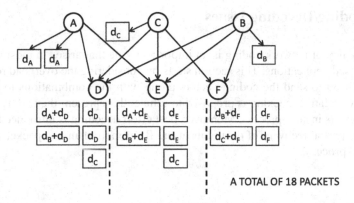

Fig. 3.3 Protecting against three losses

u needs to participate in $\frac{e+1}{2}$ F-processes. On the other hand, if e is even, node u needs to participate in $\lceil\frac{e+1}{2}\rceil$. A better alternative is to let u participate in $\lfloor\frac{e+1}{2}\rfloor$ and send an additional packet containing only d_u with $NRC = 1$, which prevents this particular packet from being combined with others along the way to the BS. The second option is better since it produces one less packet. Finally, please note that participating means that a node can either initiate an F-process as a Type 1 node, or encode its data unit (with another packet from an F-process) as a Type 2 node.

To control the number of participations of a source for a certain data unit, assume that a source keeps a counter for each generated data unit to countdown the number of remaining participations. That is, in some node u, a counter is initialized to $e + 1$ for every generated data unit d_u. The counter is decreased by two every time the source initiates an F-process by sending two copies of d_u, or every time the source participates in an F-process by sending d_u and $d_u + d_p$. Of course, when the counter reaches zero no further participations are needed. An example is shown in Fig. 3.3, where three packet losses need to be tolerated for each of the data units, i.e., $e = 3$. There are three Type 1 nodes (A, B and C), and three Type 2 nodes (D, E and F), each of which has a single data unit to be sent. Consider the packets going through nodes D, E and F, where each of these nodes forward traffic from more than one F-process, and assume that the paths for the remaining packets from A, B, and C do not pass through Type 2 nodes. For every data unit a counter is initialized to 4, i.e., each Type 1 node can *initiate* 2 F-processes and each Type 2 node can *participate* in 2 F-processes. For example consider node A, the two packets sent from A to nodes D and E represent one process initiated by A, and the remaining two packets represent the other process initiated by A after which the counter will be 0 for d_a. Nodes D and E receive packets from all the three Type 1 nodes, but since they can only participate in two processes, they choose to participate in the processes initiated by nodes A and B. Therefore, when nodes D and E receive the packet from C they act as Type 3 nodes and just forwards the packet as is. As shown in Fig. 3.3, this operation results in a total of 18 packets to protect the data units from 6 sources against 3 losses.

3.3   Coding/Decoding Issues

The efficiency of network coding is an important issue that arises in most network coding-based applications. This section starts by considering the overhead resulting from the need to send the coding vectors along with the combinations to the BS. The idea of relative indexing is proposed to efficiently represent the coding vectors for the packets in an F-process. Moreover, best effort decoding is introduced, which allows the partial recovery of data units at the BS if more than one packet was lost from an F-process.

3.3.1   Relative Indexing for Efficient Encoding

Successful decoding at the BS requires each packet to contain a coding vector, which in turn, contains the coding coefficients of the combined data units. So from the coding vector, it can be known what data units are combined in a packet, and by what coefficient each data unit is multiplied. In binary network coding, the coding coefficients can be either 0 or 1. In multicast connections, the length of the coding vector is determined by the minimum max-flow h between the single source and any of the terminals [1]. In such a case, since all data units are originating from the same source, the index of each data unit in the coding vector can be assigned by the source, where position i in the vector is reserved for d_i. That is, if binary network coding is used, a 1 in index i means that d_i is present in the combination, and a 0 means that it is not. In the case of convergecast, the coding is distributed on multiple sources, hence, the structure of the coding vector cannot be determined by a single source. Therefore, if the WSN has N nodes, the coding vector should be of length N since any node in a WSN is a possible source at some point in time. However, note that only h nodes can participate in any F-process, which means that only h bits will be used in the N-bit coding vectors of the packets from a certain F-process. Obviously, this is an inefficient way to represent the coding vectors, especially if the network has a large number of nodes (in hundreds or even in thousands).

   This section introduces the idea of relative indexing to enable the source nodes in an F-process to use an $(h-1)$-bit coding vector. Relative indexing relies on the *NRC* value in addition to a new control bit, which is referred to as I, that will be added to the packet header. The proposed indexing process does not use index 0 in a coding vector, and it always assigns index 1 to the initiator of an F-process. The indexing process recursively divides the remaining indices between the remaining sources. Assume that both packets generated by an initiator node A will be combined with data units from two other sources B and C. The question is if A is given index 1 in the $(h-1)$-bit coding vector, how can B and C (and any other Type 2 node in the process) choose distinct indices using the values of *NRC* and control bit I?

   As mentioned earlier index 1 will be given to the initiator, while index 0 will not be used. This leaves $(h-2)$ indices which will be recursively divided into halves.

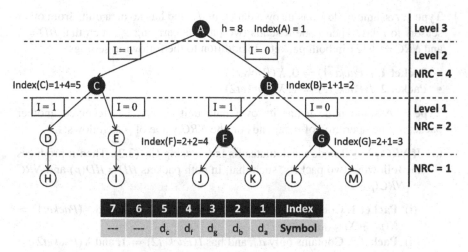

Fig. 3.4 Relative indexing, for $h = 8$

Consider the pair of Type 2 nodes in the first coding level after A (i.e., B and C), where each of them receives a packet with $NRC = \frac{h}{2}$. To allow each node to pick a different index, the control bit I takes different values. Specifically, I will be set to 0 in the packet sent to B, and it will be set to 1 in the packet sent to C. Node B is given the first index in the first half, and node C is given the first index in the second half of the remaining $h - 2$ indices. Half of $h - 2$ is $NRC - 1$ (since $h = 2 * NRC$), which means that relative to index 1 the first half begins at the next position **after** *position 1* [i.e., 1+(1)], and the second half begins **after** the next $NRC - 1$ positions after *position 1* (i.e., 1+($NRC - 1$)+1). This is shown in Fig. 3.4, where h is 8, the data unit from node B will be given index $1 + (1) = 2$, and the data unit from node C will be given the index $1 + (NRC - 1 + 1) = 1 + NRC = 5$. In the second iteration the process continues but with reference to position $1 + (1)$ for the nodes descending from B, and with reference to position $1 + (NRC - 1) + 1$ for the nodes descending from C. Figure 3.4 shows how the indexing process works for the descendants of B, where the NRC is set to 2 in the packets received by nodes F and G, resulting in G taking index $2+(1) = 3$ and F taking index $2 + (2 - 1 + 1) = 4$.

To generalize for other levels, let $X(p)$ be the coding vector of the combination carried in p, and let X_i be an $(h - 1)$-bit vector with 1 in the ith position and zeros otherwise. In addition, let $I(p)$ and $NRC(p)$ denote the value of bit I and the NRC field in p respectively. If a node u decides to participate in the F-process of some packet p, it calculates its index according to the following equation:

$$Index(u) = \max_{i:X_i \wedge X(p)=X_i} i + I(p) * (NRC(p) - 1) + 1 \tag{3.2}$$

where the values for $I(p)$ and $NRC(p)$ are set according to following operation (we only redefine the operation for Type 1 and Type 2 nodes):

1. **Type 1.** Assume node u has its own data unit d_u, and has no data units from other sensors to relay. Then node u sends two packets carrying d_u, such that $IID = u$ and $NRC = \lfloor h/2 \rfloor$ in both packets. In addition to the following settings:

   * **Packet 1.** $I(Packet1) = 0, X(Packet1) = X_1$.
   * **Packet 2.** $I(Packet2) = 1, X(Packet2) = X_1$.

2. **Type 2.** Assume node u has its own data unit d_u and has received another packet p. The operation of u depends on the NRC value of p as follows:

   (a) If $NRC(p) \geq 2$ node u calculates its index using Eq. (3.2). Then it produces the following two packets, such that, in both packets $IID = IID(p)$ and $NRC = \lfloor NRC(p)/2 \rfloor$.

       (i) **Packet 1.** Contains $d_u \oplus d_p$, and has $I(Packet1) = 0$, and $X(Packet1) = X(p) \oplus X_{Index(u)}$.
       (ii) **Packet 2.** Contains only d_u, and has $I(Packet2) = 1$, and $X(Packet2) = X_{Index(u)}$.

   (b) However, if $NRC(p) < 2$ node u acts as a node of type 3, i.e., just a relay.

   Finally, it should be emphasized that relative indexing is valid only for a set of combinations having the same IID.

3.3.2 Best Effort Decoding

Section 3.1 showed how to protect the data units in an F-process against a single loss, but what if more than one loss occurs? will this render the remaining combinations unsolvable? The coding process defined in Sect. 3.1 has the nice property of producing a set of combinations that can have decodable subsets. Therefore, if more than one loss occurs, the remaining combinations of an F-process should not be discarded since a subset of the data units can be salvaged.

As the reader may have noticed, the graph induced by the nodes generating or forwarding the packets of an F-process resembles a binary tree, especially when no node receives two packets from the same F-process. Assuming e losses have occurred, the remaining combinations after the e losses will still be solvable, if the e lost combinations remove $e - 1$ data units, i.e., $1 + N_1^j - e$ equations remain in $N_1^j - (e - 1) = 1 + N_1^j - e$ unknowns. Using the binary tree representation, this condition is satisfied for any subtree rooted at a Type 2 node u that combines its data unit with a trivial combination composed of a native data unit.

Consider a subtree rooted at a Type 2 node u, which encodes its data unit with a single data unit combination containing only d_v. The number of combinations produced by this subtree is equal to the NRC value in the packet carrying d_v. In addition, the number of new data units that will be added in these combinations by the sources in the subtree (including the root node u) is $NRC - 1$. Thus there

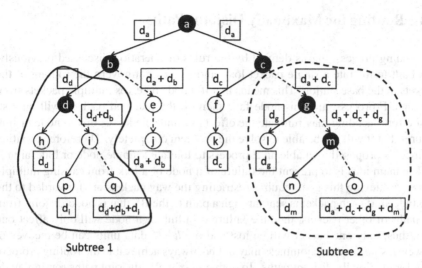

Fig. 3.5 Best effort decoding

will be *NRC* combinations in *NRC* unknowns, which can be solved. However, if the Type 2 node *u* combined its data unit with a non-trivial combination, the number of unknowns will be larger than the number of combinations. An example is shown in Fig. 3.5, where the Type 2 nodes that receive trivial combinations are the heads of all solid links, and the Type 2 nodes that receive non-trivial combinations are the heads of all dashed links. For instance the combinations from Subtree 1 are solvable since node d combines its data unit with a trivial combination, while those from Subtree 2 are not.

In reality the BS does not need to search for solvable subtrees to decode them. Rather, it can use the decoding method in Algorithm 3.1 for a set of combinations having the same *IID*, where *RCV* is the set that will contain the recovered data units, and *Z* is a set for temporary use in decoding. This way the BS will recover as many data units as possible, and will stop when no new single data units are found, hence the name best effort decoding.

Algorithm 3.1 Best Effort Decoding

1: $RCV = \phi$
2: $Z=$ All single data unit combinations.
3: **while** $Z \neq \phi$ **do**
4:   For each element in Z remove it from all combinations containing it, using bitwise XOR.
5:   $RCV = RCV \cup Z$
6:   $Z=$ All new single data unit combinations.
7: **end while**

3.4　Routing for Maximally Disjoint Paths

The coding process, that is defined by the rules of operation discussed previously, was built to tolerate a single packet loss, and it does not govern the routing of the packets to the base station. This means that if a node receives multiple packets from the same F-process, and this node fails then all the received packets will be lost. Best effort decoding may mitigate the effects of multiple losses due to node or link failures, but it will not be able to solve the problem completely. Therefore, a routing protocol is proposed to enable an F-process to tolerate a single node or link failure.

The main idea is to prevent the failure of a node or a link from causing multiple losses. Achieving this goal requires restricting the way packets are forwarded to the BS. Indeed, if each packet is sent through a path to the BS that is node-disjoint from the paths of other packets, then the failure of a link or a node will only affect one path, thus, only one packet will be lost and the $h - 1$ data units can be recovered. However, since node-disjointness may not be always achievable the routing protocol aims for maximally disjoint paths. To define maximally disjoint paths routing, node and link disjointness for an F-process need to be clarified first.

The traditional definition of node disjoint paths requires that paths not have any intermediate node (between the source and destination) in common. This definition eliminates coding opportunities since coding needs the data of two or more sources to meet and be combined at a common node. Therefore, node-disjointness needs to be redefined, for the paths carrying data from an F-process, to allow combining data units and still tolerate a single node failure. Recall that every source node (Type 1 or 2) generates and sends two packets to the BS, which means that each source needs two paths, one for each packet. The h paths are said to be node-disjoint, if the graph induced by all the nodes (except the BS) on all paths construct a binary tree. Put differently, for a certain F-process, the h paths are said to be node-disjoint if a node of Type 2 or 3 is part of at most one path from a source node (Type 1 or 2) to the BS. This definition allows sharing path segments between source nodes. For example, consider node a in Fig. 3.5, the paths from this node are $a \rightarrow c \rightarrow g \rightarrow m \rightarrow o \rightarrow BS$, and $a \rightarrow b \rightarrow e \rightarrow j \rightarrow BS$, which are totally disjoint. Now consider node c, which is a Type 2 on one of the paths from node a to the BS (i.e., node a is the predecessor of node c). Obviously, the path segment $c \rightarrow g \rightarrow m \rightarrow o \rightarrow BS$ is shared with node a. Note that the definition prevents a Type 2 node from being on more than one path from a predecessor to the BS, but it does not prevent a path from having more than one Type 2 node on the way to the BS. In fact, nodes g and m on the shared path segment between c and a are Type 2 nodes and they share part of the segment with c and a too.

Practically, the definition of node-disjointness translates to the following requirements:

1. A Type 2 node must not combine its data more than once with the packets in a certain F-process (identified by IID and a timestamp).
2. If a Type 2 node acts as a relay it can only forward one packet from a certain F-process.
3. A Type 3 node can only forward one packet from a certain F-process.

Recall that, to prevent producing dependent combinations, Type 2 nodes already do not combine their data units with more than one packet from the same F-process. What remains is to apply this restriction on relaying data as well (either by Type 2 or 3 nodes).

Note that if the paths are node-disjoint, then they can tolerate both node and link failures. However, depending on the underlaying topology, achieving node-disjointness may not be always possible. In such a case, the routing protocol should aim for link-disjointness to be able to tolerate link failures at least. Traditionally, link disjointness means that no two paths from one or more source nodes should have a link in common. As seen previously, since source nodes can share path segments to the BS, this definition of link disjointness is not suitable too. For an F-process, the paths forwarding the packets generated by an F-process are said to be link-disjoint, if no node sends two packets from the same F-process to the same downstream node.

Routing for maximally disjoint paths means that the routing protocol should try to achieve node-disjointness first. However, if this is not possible, it should try to achieve link-disjointness. Finally, if even link disjointness is not possible, the routing protocol should try to share as few links as possible.

For the routing protocol to work, the underlying WSN needs to be organized into levels or rings around the BS, as assumed at the beginning of the chapter. Initially, a node (say v) has its level l_v set to a very large value. Arranging the network into levels can be done through a simple process initiated by the BS. The BS starts by broadcasting a control packet containing a field called *hop_count*, which is initialized to 1. After that, if a node v receives this control packet with $hop\_count \geq l_v$, it discards the packet. Otherwise, it makes $l_v = hop\_count$, increases the *hop_count* by 1 and rebroadcasts the control packet.

WSNs are usually densely deployed. Therefore, during the process of organizing the network into levels, a node, v, will most likely receive more than one packet with the same minimum *hop_count* from nodes in $l_v - 1$. Each node stores the IDs of these neighbors in a list called *next_hop_list*. If a node needs to forward a data packet to the BS, it only needs to send the packet to one of the nodes stored in the *next_hop_list*. When a node receives a packet it just checks the level of the sender, if the sender is farther from the BS (in a higher level), then it chooses a node from its own *next_hop_list* and rebroadcasts the packet. Otherwise, if the sender is closer to the BS, the heard packet is discarded. This protocol would suffice if we are talking about packet loss only; a Type 1 node can include in the packet two nodes from its *next_hop_list* to generate the needed redundancy; a Type 2 node includes the id of a different neighbor for each packet; and a Type 3 node just picks one neighbor from the *next_hop_list*. However, as discussed earlier, to tolerate node failures, each node must be limited to forward only one packet at most from each F-process. To eliminate extra routing overhead, we propose to exploit the protocol used in MAC layer. So the proposed routing mechanism does not deal with channel access, we only assume that there is some CSMA/CA MAC layer protocol that does this for us, and we are just tweaking it to eliminate any extra routing overhead.

The operation of Type 3 nodes is now defined to produce maximally disjoint paths (the operation for Type 1 and 2 nodes is similar, but takes into account two packets instead of one). To establish node-disjoint paths from the multiple sources in the same F-process, a Type 3 node must not forward more than one packet from the same F-process (i.e., with the same *IID*). To do this, each node needs to know for which F-processes it had forwarded packets before. Therefore, each node stores the *IID* of each packet it forwards in a list called the *IID_list*. If node v is forwarding a packet p with $IID(p)$, it must select a neighbor in the *next_hop_list* that does not have $IID(p)$ in its *IID_list*. This can be accomplished during the RTS/CTS negotiation (or a separate handshake if RTS/CTS is not used), where the *IID* of packet p is piggybacked in the RTS message, and a next-hop node in $l_v - 1$ can reply with a CTS if it does not have $IID(p)$ in its *IID_list*. This forwarding scheme guarantees node-disjointness as defined before.

If node v does not find a neighboring node in $l_v - 1$ that does not have $IID(p)$ in its list, then node-disjointness will not be possible, but edge-disjointness may still be possible. Node v knows that all nodes in $l_v - 1$ have $IID(p)$ in their lists if it does not receive a CTS. Therefore, to achieve edge-disjointness a node in a $l_v - 1$ needs to be forced to send a CTS even if it had $IID(p)$ in its *IID_list*. A new field is assumed to be added to the RTS message, which is basically a flag (bit) called *"Force forward"* or simply *FF*. This flag should only be set in the second RTS, if a node does not receive a CTS for its first RTS. If a neighboring node in $l_v - 1$ hears an RTS with $FF = 1$ it replies with a CTS regardless if it has $IID(p)$ or not. The operation of a node upon receiving a packet is described in Algorithm 3.2, and its operation when sending a packet is described in Algorithm 3.3.

Algorithm 3.2 Operation of Node v: Receiving Packet p

```
 1: if (p is a control packet) then
 2:     if (hop_count > l_v) then
 3:         Discard p
 4:     else if (hop_count < l_v) then
 5:         l_v = hop_count
 6:         Clear next_hop_list
 7:         Rebroadcast p
 8:     else
 9:         //hop_count = l_v
10:         Add sender of p to next_hop_list
11:         Rebroadcast p
12:     end if
13: else if (p is RTS received from level l_v + 1) then
14:     if (if IID(p) ∉ IID_list) then
15:         Send CTS
16:     else if (FF(RTS) == 1) then
17:         Send CTS
18:     end if
19: else
20:     //p is a data packet
21:     Store IID(p) in IID_list
22:     Send p downstream
23: end if
```

Algorithm 3.3 Operation of Node v: Sending Packet p

1: **if** (p is a control packet) **then**
2:    Broadcast p
3: **else**
4:    //p is a data packet
5:    Piggyback $IID(p)$ on RTS
6:    **if** (CTS is received from node u before timeout) **then**
7:        Send p to node u
8:    **else**
9:        $FF(RTS) = 1$
10:      Wait for CTS
11:      Upon receiving CTS from node u, send p to u
12:    **end if**
13: **end if**

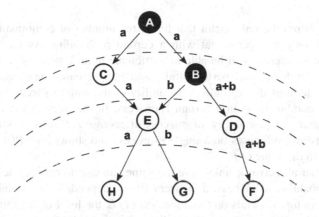

Fig. 3.6 The figure shows how the combinations created in an F-process initiated by node A are forwarded. Node C cannot find a downstream neighbor that did not forward a packet with IID = A. Therefore, it force forwards data unit a to node E, **which results in paths that are not node-disjoint**. Note that if another neighbor to node C is found the paths would be node-disjoint. Note also that if there were only two nodes in $l_E - 1$, then E will force forward one of the data units (either a or b), and the resulting paths will be neither link nor node-disjoint

It is easy to see that by using the described routing protocol, all paths from all sources will be node-disjoint unless some node force forwards a packet. Also, the paths will be link-disjoint unless a node forwards two packet to the same next hop. By using such a simple routing protocol, we can say that the F-process can tolerate at most one node (or link) failure if the paths from all sources are node (or link) disjoint. An example is shown in Fig. 3.6, where an F-process is initiated by node A, and node B is the only downstream Type 2 node. In the example, node C force forwards the packet carrying data unit a because there are only two nodes in $l_C - 1$. After that, no force forwarding occurs, which produces edge-disjoint paths. Note that if another neighbor to node C is found in $l_C - 1$, then the paths would be node-disjoint. However, note that if there were only two nodes in $l_E - 1$, then E will force forward one of the data units a or b, and the resulting paths will not even be edge-disjoint. It should be noted that if the paths are totally disjoint, then the failure of a node can cause at most one combination to be lost even if it is a Type 2 node. Let v be a Type 2 node that receives packet p from some F-process, then if v fails before sending its two packets containing $d_v \oplus d_p$ and d_v, the failure of v will only cause d_p to be lost.

At the beginning a dense WSN was assumed, as it is usually the case in most WSNs. However, in a worst case scenario where there is only one path from a node to the BS and all the nodes on the path are Type 2 nodes, then all the packets in the F-process will be sent (force forwarded) on the same path.

3.5 Selecting Parameter h

The min-cut is not the only factor that limits the number of combinations h. Any network link may be operational with a certain probability. As the value of h gets larger more sources will be able to participate in an F-process, which in turn increases the number of used network links, and thus increases the chance of having multiple link failures that can affect the ability of the sink to successfully decode the received combinations in a certain F-process. In this section the relationship between h and the probability of successful recovery $P(rcv)$ is studied. The following discussion focuses on a single F-process and shows how $P(rcv)$ changes with respect to the value of h.

Assume that all network links have the same success probability denoted by q. Then the probability of successful recovery $P(rcv)$ depends on the number of used links, which in turn depends on three factors: (1) L, the level of the initiator node, (2) $h - 1$, the number of sources that will participate in the F-process, and (3) the way the sources are distributed in the levels below the initiator node, or equivalently the resulting topology of the tree. The probability of successful recovery given some known topology, Tp_1, with k_1 links, is equal to the probability that at most one loss occurs in Tp_1, i.e.,

$$P(rcv|Tp_1) = P(\text{no loss in } Tp_1) + P(1 \text{ loss in } Tp_1)$$

$$= q^{k_1} + k_1(1 - q)q^{k_1 - 1}$$

Therefore, $P(rcv)$ can be computed as follows:

$$P(rcv) = \sum_{\forall Tp_i} P(rcv|Tp_i)P(Tp_i)$$

Since all network nodes generate data units at the same rate, i.e., the probability of being a source is the same for all nodes, then all possible topologies can occur with the same probability. Therefore, the probability of successful recovery can be written as a function of L and $h - 1$ as follows:

$$P(rcv) = P_{rcv}(L, h - 1) = \frac{\sum_{\forall Tp_i} P(rcv|Tp_i)}{\#\text{possible topologies}}$$

$$= \frac{\sum_{\forall Tp_i} (P(\text{no loss in } Tp_i) + P(1 \text{ loss in } Tp_i))}{\#\text{possible topologies}}$$

$$= \frac{P_0(L, h - 1) + P_1(L, h - 1)}{\#\text{possible topologies}}$$

$P_0(L, h - 1)$ is the sum of probabilities of no loss in all the possible topologies starting at level L with $h - 1$ sources, and $P_1(L, h - 1)$ is the sum of probabilities of exactly 1 loss in all the possible topologies starting at level L with $h - 1$ sources. P_0 and P_1 are evaluated for trees starting at nodes of Types 1 or 2. Following the rules of operation, the initiator node will send two packets containing copies of its data unit to the BS. As these packets are forwarded, a Type 2 node may participate and code its data with the data unit received from one of these packets, which also produces two packets according to the rules of operation that may also trigger other downstream Type 2 nodes to participate. For the purpose of computing $P(rcv)$, the contents of the packets forwarded from Type 1 or Type 2 nodes are ignored, and only the number of transmissions or produced packets is considered. This way an F-process (initiated by a Type 1 node) can be viewed as a recursive process that repeats itself (in some lower level, and with a fewer number of sources) on every downstream source (Type 2 node). This allows the calculation of $P_0(L, h - 1)$ and $P_1(L, h - 1)$ using recursive formulas, but first the following definitions are needed:

- let $\pi(\lambda, \alpha)$ be the probability that there is no loss in the branch starting at level λ, with α sources downstream.
- let $\bar{\pi}(\lambda, \alpha)$ be the probability that there is a loss in the branch starting at level λ, with α sources downstream.
- let $\mu(\lambda, \alpha)$ be the probability that a loss will occur downstream, given that we are starting at level λ with α sources and no loss has occurred yet.
- let $\nu(\lambda, \alpha)$ be the probability that no loss will occur downstream, given that we are starting at level λ with α sources and a loss has already occurred.
- let $\rho(\lambda, \alpha)$ be the probability that no loss will occur downstream, and the root is not necessarily a source. This is similar to $P_0(\lambda, \alpha)$, but branching may not happen at level λ.

Notice that $\pi(.,.)$ and $\bar{\pi}(.,.)$ are evaluated at nodes of Type 3, since they start with branches, or links. Also note that $\mu(.,.)$ and $\nu(.,.)$ start at nodes on the tree which may be of any type, i.e., they have degrees 1 or 2.

Using these definitions P_0 and P_1 can be written as follows:

$$P_0(L, h - 1) =$$

$$\sum_{\alpha_1=0}^{\lfloor \frac{h-1}{2} \rfloor} \quad \sum_{\forall \{\alpha_2 : \alpha_1+\alpha_2=h-2, \, 0\leq\alpha_2\leq\lfloor\frac{h-1}{2}\rfloor\}} \pi(L, \alpha_1)\pi(L, \alpha_2) \qquad (3.3)$$

$$P_1(L, h - 1) =$$

$$\sum_{\alpha_1=0}^{\lfloor \frac{h-1}{2} \rfloor} \quad \sum_{\forall \{\alpha_2 : \alpha_1+\alpha_2=h-2, \, 0\leq\alpha_2\leq\lfloor\frac{h-1}{2}\rfloor\}} \pi(L, \alpha_1)\bar{\pi}(L, \alpha_2) \qquad (3.4)$$

$$+ \sum_{\alpha_1=0}^{\lfloor \frac{h-1}{2} \rfloor} \quad \sum_{\forall \{\alpha_2 : \alpha_1+\alpha_2=h-2, \, 0\leq\alpha_2\leq\lfloor\frac{h-1}{2}\rfloor\}} \bar{\pi}(L, \alpha_1)\pi(L, \alpha_2)$$

where,

$$\pi(L, \alpha) = q.\rho(L - 1, \alpha) \qquad (3.5)$$

with $\pi(1, 0) = q$, and $\pi(1, \alpha \geq 1) = 0$ because there are no sources below level 1.

$$\bar{\pi}(L, \alpha) = q.\mu(L - 1, \alpha) + (1 - q).\nu(L - 1, \alpha) \qquad (3.6)$$

with $\bar{\pi}(1, 0) = 1 - q$, and $\bar{\pi}(1, \alpha \geq 1) = 0$.

$$\mu(L, \alpha) = q.\mu(L - 1, \alpha) + (1 - q).\nu(L - 1, \alpha) + P_1(L, \alpha) \qquad (3.7)$$

with $\mu(1, 0) = 1 - q$, $\mu(1, 1) = 2q(1 - q)$, and $\mu(1, \alpha > 1) = 0$.

$$\nu(L, \alpha) = \nu(L - 1, \alpha) + P_0(L, \alpha) \qquad (3.8)$$

with $\nu(1, 0) = q$, $\nu(1, 1) = q^2$, and $\nu(1, \alpha > 1) = 0$.

$$\rho(L, \alpha) = P_0(L, \alpha) + q.\rho(L - 1, \alpha) \qquad (3.9)$$

with $\rho(1, 0) = q$, $\rho(1, 1) = q^2$, and $\rho(1, \alpha > 1) = 0$.

Finally, the total number of possible topologies needs to be calculated to be able to calculate $P_{rcv}(L, h-1)$. To do this, $P_0(L, h-1)$ can be used. Since P_0 computes the sum of the probability of no loss in all possible topologies setting $q = 1$ will make the probability of no loss equals 1 for each topology, and P_0 will result in counting the total number of possible topologies. That is, $P_{rcv}(L, h - 1)$ can be calculated as follows:

$$P_{rcv}(L, h - 1) = \frac{P_0(L, h - 1) + P_1(L, h - 1)}{P_0(L, h - 1)|_{q=1}} \qquad (3.10)$$

3.5.1   Evaluating P(rcv)

The probability of successful recovery $P(rcv)$ is plotted against the total number of sources in an F-process starting at different levels, for $q = 0.99$, $q = 0.999$, and $q = 0.9999$. The results are shown in Figs. 3.7, 3.8, and 3.9. As expected, for some level L, increasing the number of participating sources in an F-process will reduce $P(rcv)$ because this increases the total number of used links. Therefore, depending on the desired probability of successful recovery, nodes in levels that are far away from the BS may want to just use duplication, or equivalently let $h = 2$ to prevent downstream Type 2 nodes from participating. Also, for a fixed number of sources, the figure shows that the probability of successful recovery increases as the initiator node gets closer to the BS (i.e., as L decreases). This is expected since a closer initiator means shorter paths and less links, hence, better $P(rcv)$.

As shown in the figures, closer levels (e.g., 3 or 4), cannot be evaluated for a large number of sources since the total number of source nodes is limited by $2^L - 1$. Finally, it should be noted that these calculations for $P(rcv)$ are only valid if the data units from all sources in an F-process are to be recovered. This is because (as we

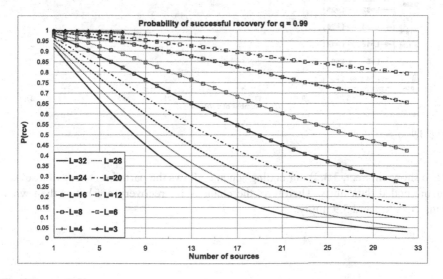

Fig. 3.7 $q = 0.99$

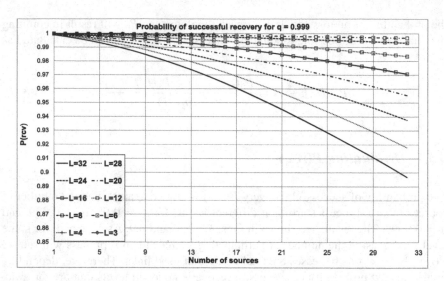

Fig. 3.8 $q = 0.999$

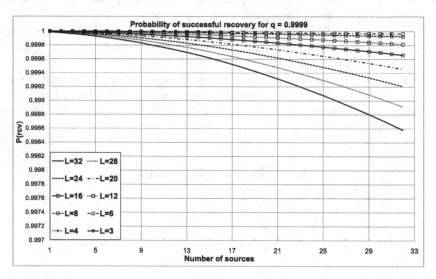

Fig. 3.9 $q = 0.9999$

have shown in Sect. 3.3.2) our coding scheme allows the partial recovery of a subset of the coded data units. Therefore, the chances for recovery are better than what Figs. 3.7, 3.8, and 3.9 imply.

3.6 Simulation Results

The performance of the proposed scheme is evaluated in terms of produced overhead using simulation on TOSSIM. The effects of the following three parameters were studied in the experiments:

1. The number of packets generated by each sensor node, denoted by S.
2. The number of nodes in the network, denoted by N.
3. The maximum number of produced combinations in an F-process, denoted h.

As a reference, the results are compared to a theoretical lower bound, which represents the best way in which packets can be combined. The best case occurs when every $h-1$ nodes belong to a certain F-process in which all of them are either Type 1 or Type 2 nodes. That is, there will be $\frac{N}{h-1}$ groups of nodes each of which produces a total of hS packets. This gives the following lower bound on the total number of packets produced:

$$(\frac{h}{h-1})NS \tag{3.11}$$

Three scenarios were considered, where in all of these scenarios h took the values $\{2, 4, 8, 16\}$. Note that when $h = 2$ no coding will take place and the proposed scheme will be equivalent to traditional duplication based redundancy. The first scenario is a 7×7 grid network (i.e., N = 49) where each node produced 20 packets, the result is shown in Fig. 3.10, where at $h = 16$ the duplication overhead is reduced by 42 %. The second is also a 7×7 grid, but each node produced 50 packets to see how increasing the number of produced packets affects the performance. Figure 3.11 shows that by increasing the number generated packets the performance

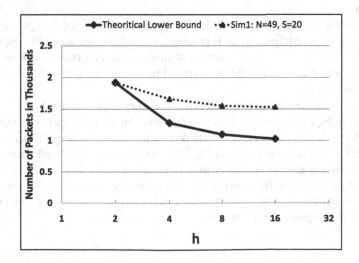

Fig. 3.10 Simulation results for N = 49 and S = 20

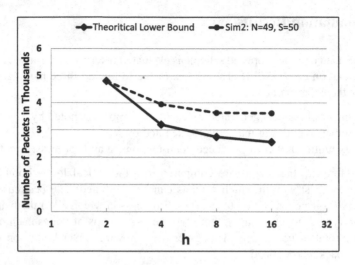

Fig. 3.11 Simulation results for N = 49 and S = 50

gets better and the duplication overhead is decreased by 50 % at $h = 16$. Finally, the size is increased to a 10×10 grid network (i.e., N = 100) and the number of produced packets is set to 50, to see how the performance is affected by increasing the network size in addition to the number of packets. Again, the results assure that the performance gets better with increasing the network size and generated packets, where at $h = 16$ the duplication overhead is reduced by 72 %. It is clear that the third scenario is the closest to the lower bound, since the chances for combining packets are enhanced as the network and/or number of generated packets grows larger. Although these scenarios are relatively small in size, they clearly illustrate that the coding-based approach is scalable and performs better as the network size increases.

An important thing to note here is that there is no need for a large number of Type 2 nodes to participate in an F-process to significantly reduce the duplication overhead. Note that the largest decrement in the overhead occurs at $h = 4$, and that the reduction in the duplication overhead becomes negligible after the first few participating sources.

To see how the probability of successful recovery ($P(rcv)$) affects the overhead, the results in Fig. 3.12 are modified to reflect the change in $P(rcv)$. The number of packets for each point in Fig. 3.12 is divided by the corresponding $P(rcv)$ (at $h - 1$) from Fig. 3.7. The result is shown in Fig. 3.13. It is clear from the figure that network coding might not always be better than duplication especially if an F-process is initiated at a level that is far from the BS. As previously mentioned, the figure shows that a good reduction in the overhead can be achieved by just letting a small number of sources participate in an F-process, e.g., 3 (at h = 4).

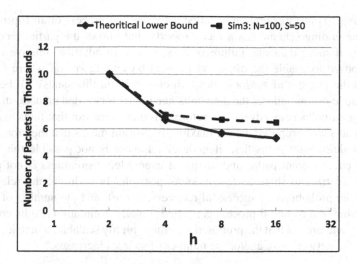

Fig. 3.12 Simulation results for N = 100 and S = 50

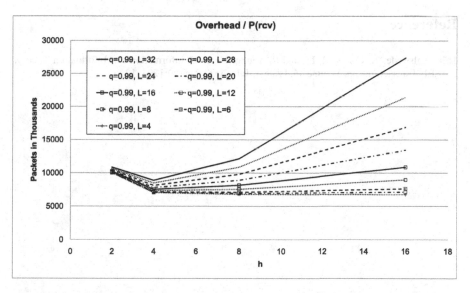

Fig. 3.13 The figure shows how the produced overhead changes with respect to the initiator node level (L), and the number of participating sources

3.7 Chapter Summary

This chapter introduced a distributed network coding-based protection scheme that defines the operation for each node relying on its local information only. The proposed protocol uses binary network coding to allow a set of k source nodes to collaboratively produce $k + 1$ linear combinations such that any k combinations are solvable. This collaboration is called a forwarding process or an F-process for short. The coding scheme has a nice property that allows the partial recovery of data units if more than one failure or loss occurs. In addition, relative indexing was proposed to enable the use of an $(h - 1)$-bit coding vector, where $h << N$ and N is the number of sensor nodes (which can be in thousands). To be able to tolerate node or link failures the data units need to be forwarded on node-disjoint or edge-disjoint paths respectively. Therefore, a routing protocol that finds maximally disjoint paths was proposed, where maximally disjoint means that the protocol tries to find node-disjoint paths first, if node-disjointness is not possible the protocol aims for edge-disjoint paths, and finally, if even edge-disjointness is not possible the protocol tries to share as few links as possible. In addition, the relationship between the probability of successful recovery, $P(rcv)$, and the number of sources participating in a certain F-process is studied. Finally, simulation results confirmed the theory and proved that the proposed scheme is highly scalable, where it performs better as the network size and/or the number of sources increases.

Reference

1. R. Ahlswede, N. Cai, S. R. Li, and R. Yeung. Network information ow. Information Theory, IEEE Transactions on. Volume 46, Issue 4, July 2000 Page(s):1204–1216.

Chapter 4
Transmissions Scheduling in Network Coding-Based Resilient WSNs

Network coding is a generalization of routing and allows network nodes to combine data units from different packets instead of forwarding the packets as is. This means that a node needs to correctly receive and decode the packets first to obtain their data content. After that, the data can be combined and sent from the node. In this case, the coding process takes place in the electronic domain after obtaining the actual bit representations of the data in the packets. This type of coding is known as Digital Network Coding (DNC), and it is the dominant coding strategy. Alternatively, Analog Network Coding (ANC) [1] utilizes interference of packets and carries the coding (combining of data) process in the electromagnetic domain at the physical layer. That is, it does not require the successful reception of the packets or their actual bit representations.

This chapter augments and complements the proposed scheme in Chap. 2. Specifically, this chapter studies the problem of scheduling the transmissions of packets from the source nodes in U to the coding nodes in L under both DNC and ANC. Bounds on the number of needed time slots for both coding strategies are obtained. Also, this chapter considers some special cases for ANC when the graph induced by U and L has certain properties. To compare the performance of DNC and ANC, a greedy scheduling algorithm is used, and the results show that ANC outperforms DNC by a factor of 2 approximately.

4.1 Introduction

To clarify the difference between DNC and ANC, consider the example in Fig. 4.1, which shows a two-way relay channel where there are two terminal nodes (A and B) that need to exchange their packets through a relay node (R). This exchange consumes four time slots as clarified in the figure. In digital network coding, the packets from sources need to be received correctly at the physical layer, to get

© The Author(s) 2015
O.M. Al-Kofahi, A.E. Kamal, *Resilient Wireless Sensor Networks*, SpringerBriefs in Electrical and Computer Engineering, DOI 10.1007/978-3-319-23965-1_4

Fig. 4.1 No coding

Fig. 4.2 Digital network coding

Fig. 4.3 Analog network coding

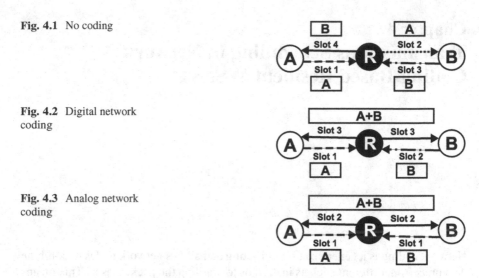

their actual bit representation, which can then be combined at higher layers. This requires each packet to be received separately, otherwise, a collision will occur. DNC can enhance on the performance in the previous example, by scheduling the transmissions of terminal nodes first, then allow the relay to combine the received data units and send the result to the terminal nodes. A terminal node can retrieve the data destined to it by subtracting its own data unit from the received combination. This reduces the needed time slots to three as shown in Fig. 4.2. Analog network coding was proposed in [1] to enhance the capacity of the two-way relay channel. Unlike DNC, ANC does not require the reception of each packet alone, rather, it benefits from collisions between packets to enhance throughput. In the two-way relay channel, if the terminal nodes transmit together, the relay node receives their added analog signals, and then forwards the result to the two terminal nodes. After that, each terminal node can recover the data destined to it by first regenerating the physical representation of its own data unit, and then subtracting it from the combined signal. That is, with ANC, two packets can be exchanged through the relay node and received by the terminals in only two time slots as shown in Fig. 4.3.

Note that with many-to-one flows, all the data units and the sum of the analog signals will be forwarded to a single common destination (the Sink or Base Station), through multihop wireless communication. When receiving all this information the sink will be able to decode the data units. This is illustrated in Fig. 4.4 that shows two source nodes A and B, and their one-hop neighboring coding nodes (the unlabeled black nodes). The edges represent the available wireless links between the nodes. Using ANC, both sources transmit in the same time slot, which results in the combinations shown in the figure at the coding nodes. Finally, note that ANC is only used by the coding nodes to create the combinations, and it is not used in the transmissions from the coding nodes to the sink.

Fig. 4.4 Protecting two data
units that are destined to node
T against a single link failure

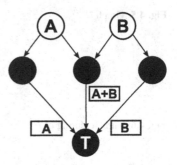

4.2 Scheduling Based on Digital Network Coding

As mentioned earlier, digital network coding needs to know the actual bit representation of the received data units to be able to perform coding correctly. Therefore, if a coding node in L is to combine the data units from source nodes u and v in U, then each of these sources needs to transmit in a separate time slot. This defines the *feasibility condition* of a DNC-based schedule, and it must be guaranteed in all the time slots of any DNC-based schedule. Hence, a conflict exists between source nodes u and v if they have a common neighboring coding node in L, meaning, they cannot be scheduled in the same time slot.

DNC-based scheduling can be shown to be NP-Complete by a direct reduction from the vertex-coloring problem, where given a graph $H = (V', E')$ a color must be assigned to every vertex in V', such that, no two adjacent vertices have the same color [2]. H is called a conflict graph because, in the vertex-coloring problem, a link (u, v) represents a conflict between two nodes u and v. Given any conflict graph H, it can be reduced to an instance of DNC-based scheduling through the following steps:

- The nodes in V' are mapped to the nodes in U.
- An edge (u, v) in H is transformed to a node n_{uv} that belongs to L, and is connected to nodes u and v in U.
- For each node in U add a neighbor in L and connect them together to satisfy the connectivity requirements in Chap. 2 (Sect. 2.2.2).

Thus an optimal and feasible schedule of k time slots exists for G if and only if H is k-chromatic, i.e., each color in H represents a time slot in G and vice versa. This mapping is not the only possible mapping from the vertex coloring problem to the DNC-based scheduling problem, however, this does not invalidate the reduction. Note that using this transformation all the nodes in L will have degree 2. Nevertheless, this does not mean that a conflict graph for G can only be constructed if all of the L nodes in G have degree 2. For example consider the graph in Figs. 4.5 and 4.6, applying the transformation stated above will result in the graph shown in Fig. 4.7. However, this is not the only graph that has a conflict graph similar to H in Fig. 4.5. For instance the graph shown in Fig. 4.8 will have exactly the same conflict graph H, and thus solving the vertex-coloring on H provides a feasible

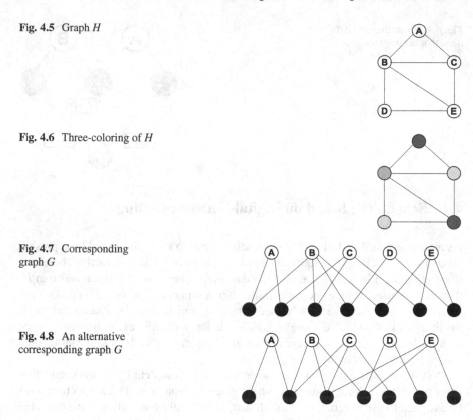

Fig. 4.5 Graph H

Fig. 4.6 Three-coloring of H

Fig. 4.7 Corresponding graph G

Fig. 4.8 An alternative corresponding graph G

and optimal schedule for both the graphs in Figs. 4.7 and 4.8. Reciprocally, finding a feasible schedule for either of the graphs in Figs. 4.7 or 4.8 will solve the coloring problem on H. It is worth mentioning that although in the coming discussion all the nodes of L in the used examples have degree 2, the analysis is valid and applies for any graph G.

This relation with the vertex-coloring problem is used to bound the number of needed time slots for a DNC-based schedule. Let Δ_L be the maximum node degree in L in G, let T_{schd} be the minimum number of time slots needed for a schedule (or equivalently the number of colors in H), and let Δ_H be the maximum node degree in the conflict graph H corresponding to G. We have the following bounds on the number of needed time slots.

$$\Delta_L \leq T_{schd} \leq \Delta_H$$

The lower bound follows since no two sources in U can be scheduled together if they have a common neighbor in L. The upper bound follows from Brook's theorem [2], which says that *"for a connected graph H that is neither a full graph nor an odd cycle the chromatic number $\chi(H)$ is less than or equal to the maximum node degree Δ in H"*. Note that if H is a full graph or an odd cycle then $\chi(H) = \Delta + 1$.

Fig. 4.9 Coding tree on graph G

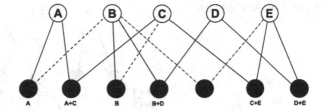

The minimum number of needed colors for H in Fig. 4.5 is three, and the solution is shown in Fig. 4.6. This indicates that an optimal schedule for the transmissions of the sources in G needs three time slots. For this example note that $\Delta_L = 2$, $T_{schd} = 3$, and $\Delta_H = 4$. Finally, one should note that scheduling and coding are two independent processes. This is because a coding tree (Sect. 2.4) can be constructed to decide the combinations after the transmissions schedule is finished. One possible coding tree for the graph G in Fig. 4.7 is shown in Fig. 4.9. The solid links represent the links on the tree, which will decide on the combinations as shown.

4.3 Scheduling Based on Analog Network Coding

Analog network coding allows a coding node in L to receive from two source nodes in U simultaneously, which means that an ANC-based schedule can have two sources with a common neighboring coding node in the same time slot. That is, the feasibility condition for ANC-based scheduling limits a coding node in L to hear from at most two sources in the same time slot. This is less restricting than the feasibility condition for DNC-based scheduling, where a node in L is allowed to hear from only one source in a time slot. Therefore, the lower bound on the minimum number of needed time slots, T_{schd}, is as follows:

$$\lceil \frac{\Delta_L}{2} \rceil \le T_{schd}$$

The vertex coloring problem cannot be easily mapped to ANC-based scheduling. However, Δ_H can still serve as a pessimistic upper bound.

It must be emphasized that, unlike in DNC-based scheduling, the coding of data units depends on the scheduling of the sources transmissions. This is because if, in a certain time slot, two sources A and B transmit to a common coding node, the sum of the transmitted analog signals will be received at the coding node, which determines the combination created at that node. This necessitates another feasibility requirement in the case of ANC-based scheduling to allow decoding at the sink. Basically, every time slot in an ANC-based schedule must have at least two leaf combinations. Figure 4.10 shows a feasible schedule based on analog network coding that takes two time slots for graph G shown previously

Fig. 4.10 ANC schedule

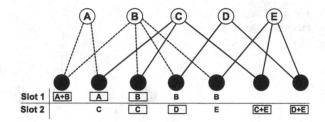

in Fig. 4.7. In the first time slot, sources A and B transmit and produce four leaf combinations. The remaining sources transmit in the second slot and produce four leaf combinations also.

If duplicate or excess combinations are created by the transmissions in a time slot of an ANC-based schedule, these duplicates need to be eliminated to reduce the number of packets sent to the destination. This can be done by finding a coding tree for the sources in each time slot and then discarding the combinations created at nodes not on the tree. For example in Fig. 4.10 we only need the combinations inside the boxes, and all other combinations are discarded.

To summarize, in a feasible ANC-based schedule the following conditions must hold:

1. In any time slot a node in L does not receive from more than two nodes in U.
2. A set of sources that transmit in a certain time slot must produce at least two leaf combinations.

4.3.1   Special Case: When $\Delta_L = 2$

If the maximum node degree in L is two, then scheduling any set of sources in the same time slot will not violate the first feasibility condition. However, the second feasibility condition still needs to be satisfied. Since $\Delta_L = 2$, there are two cases:

1. L has at least two nodes with degree 1: in this case the second condition will also be satisfied and all sources in U can transmit in a single time slot.
2. All nodes in L have degree 2: in this case any single node in U that has at least two neighbors in L can be scheduled in one time slot, and all remaining sources in U can be scheduled in the second time slot. Scheduling the node in the first time slot creates a sufficient number of leaf combinations in the second time slot to make it feasible.

4.3.2 Special Case: When G is a Tree

If G is a tree, then satisfying the connectivity assumption (Chap. 2) requires all leaf nodes to be in L. Therefore, scheduling any set of sources in a time slot cannot violate the second feasibility condition because the subtree composed of the scheduled sources and their neighbors in L will always have at least two leaf nodes. However, the first condition still needs to be satisfied in a time slot. It can be shown that for this special case the needed number of time slots is $\lceil \frac{4_L}{2} \rceil$. This is accomplished in this section by devising an algorithm that guarantees scheduling two neighboring sources for any node in L with degree larger than 2. After that, the number of needed time slots is counted.

Consider Algorithm 4.1. In a certain time slot, the algorithm starts by considering the node in L with the highest degree (node v). Then, the algorithm marks two of v's neighbors (which are nodes in U), and deletes (temporarily) all other unmarked neighbors. Marking a neighbor means scheduling it to transmission in the current time slot, i.e., v will have two of its neighboring sources scheduled, and the rest will be ignored in the current iteration (deleted). After that, the algorithm considers other nodes in L with marked or deleted U neighbors. If a node, say u, has a marked neighbor, it can mark another one because it can receive from two sources. If u has a deleted neighbor and at least two remaining neighbors, it marks two for transmission. Finally, if u has a deleted neighbor and a single remaining neighbor it just deletes this neighbor so that it will be scheduled later. When there are no more nodes to consider, time slot x is filled with the nodes in set S. To find a feasible schedule, the process is repeated until all sources are scheduled. The following lemma shows that Algorithm 4.1 will reduce the degree of any node in L by two if it has a degree larger than two.

Lemma 4.1. *If G is a tree and Algorithm 4.1 is executed, then any node in L with a degree > 2 will have its degree reduced by two after each iteration of the algorithm.*

Proof. First of all, note that for a node $v \in L$ to have its degree reduced by two (Step 21 in Algorithm 4.1), it must have two of its neighbors in U marked for transmission. In addition, the unmarked neighbors will be deleted to guarantee that a coding node does not receive from more than two sources. After that, inside the inner while loop, node v starts reaching out to its 2-hop neighbors where a 2-hop neighbor is reached when it has a marked or a deleted U neighbor. In the next iteration of the inner while loop, the 2-hop neighbors will reach-out to the 4-hop neighbors of v, and so on. The process continues until all nodes in L are reached.

The implication can be proven by contradiction. Assume Algorithm 4.1 was used, but some node $u \in L$ with degree >2 did not have its degree reduced by two after running a single iteration of the algorithm. For this to happen, node u must have had all of its neighbors deleted in the reaching-out process, which means that all of its neighbors can be reached from node v (i.e., there are multiple paths from v to u). However, this contradicts the assumption that G is a tree. Therefore, every

---

Algorithm 4.1 ANC-Based Scheduling on a Tree

Require: Graph $G = (V, E)$, $V = \{U \cup L\}$. G is bipartite, and a tree.
Ensure: Feasible ANC-based schedule that takes $\lceil \frac{\Delta_L}{2} \rceil$ time slots.
 1: $S = \emptyset, N = \emptyset, T = \emptyset, x = 0$
 2: **while** $(|T| < |U|)$ **do**
 3:     v = Node in L with max degree.
 4:     Mark any two U neighbors of v for transmission.
 5:     $S = S \cup \{\text{Marked neighbors of } v\}$
 6:     $N = N \cup \{\text{All 2-hop neighbors of v}\}$
 7:     Delete all other U neighbors of v. //Temporarily
 8:     **while** $(N \neq \emptyset)$ **do**
 9:         Remove u from N
10:         **if** (u has a marked neighbor) **then**
11:             Mark an extra neighbor of u (if any), & delete all others.
12:         **else if** (u has a deleted neighbor & remaining degree ≥ 2) **then**
13:             Mark two neighbors of u, & delete all others neighbors.
14:         **else if** (u has a deleted neighbor & remaining degree $== 1$) **then**
15:             Delete remaining neighbor
16:         **end if**
17:         $S = S \cup \{\text{Marked neighbors of } u\}$
18:         $N = N \cup \{\text{All 2-hop neighbors of u not in } S\}$
19:     **end while**
20:     Put all nodes in S in slot number x.
21:     Remove the nodes in S and their incident edges from G
22:     $T = T \cup S$, & $S = \emptyset$
23:     $x++$
24: **end while**

---

node in L with degree >2 must have its degree reduced by two after each iteration of Algorithm 4.1 is executed.   □

It can be shown now that the number of time slots is $\lceil \frac{\Delta_L}{2} \rceil$ for this special case.

Theorem 4.1. *If the bipartite graph G is a tree, a feasible ANC-based schedule can be achieved in exactly* $\lceil \frac{\Delta_L}{2} \rceil$ *time slots.*

Proof. To prove this theorem, a direct proof is used that relies on Lemma 4.1 and the special case in Sect. 4.3.1. Basically, the proof calculates a number x which equals the number of time slots needed to reduce the max-degree in L to two. When the max degree in L is two, all nodes in U will be marked by Algorithm 4.1, and non of them will be deleted. Therefore, the total number of needed time slots is $x + 1$. From the previous lemma, the maximum degree after x time slots, d_x, is:

$$d_x = \Delta_L - 2x \tag{4.1}$$

Now let $d_x = 2$, and solve for x

$$x = \frac{\Delta_L}{2} - 1$$

Adding the last time slot for the remaining sources, and taking the ceiling of $\frac{\Delta_L}{2}$ if Δ_L was odd:

$$x = T_{schd} = \lceil \frac{\Delta_L}{2} \rceil$$

$\square$

4.3.3 Maximum Gain of ANC-Based Scheduling

For a certain graph G, the ANC-gain is defined as the ratio of the number of time slots needed to schedule the transmissions of sources using DNC to the number of time slots needed to schedule the transmissions of sources using ANC. The ANC-gain is maximized when ANC-based scheduling requires the minimum number of slots, while DNC-based scheduling requires the maximum number of slots. This happens when the following is true:

1. $\Delta_L = 2$, and there are at least two nodes in L with degree 1.
2. The conflict graph H corresponding to G, is fully connected.

The first condition guarantees that only one time slot is needed for an ANC-based schedule. While the second condition guarantees that each source will be scheduled in a separate time slot in a DNC-based schedule. That is, ANC-gain will be equal to n (the number of sources). Note that if all the nodes in L have degree two, the gain will be $\frac{n}{2}$. An example is shown in Fig. 4.11, where $\Delta_L = 2$. An ANC schedule is shown in the figure, where node A transmits in the first time slot, and all remaining nodes transmit in the second time slot. The digital network coding based schedule is better shown on the corresponding conflict graph H, where each color on H represents a time slot. This is shown in Fig. 4.12.

Although the maximum ANC-gain is theoretically possible, it is highly unlikely to be achieved in practice. This is because $\binom{n}{2}$ nodes are needed in L to get a fully connected conflict graph, while keeping $\Delta_L = 2$. However, since this scheduling problem has originated from the need to provide protection against a single failure for n data units, the largest number of coding nodes that might be needed or used is no more than $2n$.

Fig. 4.11 2-slot ANC schedule

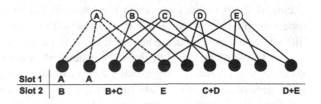

Fig. 4.12 n-slot DNC
schedule

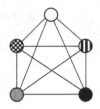

Algorithm 4.2 Scheduling Algorithm

Require: Graph $G = (V, E)$, $V = \{U \cup L\}$. G is bipartite.
Ensure: Feasible ANC-based or DNC-based schedule //depending on the feasibility conditions
       checked in Step 6
1: $S = \emptyset, F = \emptyset, N = U$
2: $Slots = 0$
3: **while** ($|S| < |U|$) **do**
4:     Select user, u, with minimum degree in N, and $u \notin S$.
5:     $S = S \cup \{u\}$
6:     $F = F \cup \{All\ sources\ infeasible\ with\ any\ source\ in\ S\}$
7:     $N = N \backslash F$
8:     **if** ($N == \emptyset$) **then**
9:        $Slots + +$
10:       $N = U, \& F = \emptyset$
11:    **end if**
12: **end while**
13: **return** $Slots$

4.4 Performance Evaluation and Comparison

In the previous sections, the examples used to compare the performance of ANC-based scheduling with DNC-based scheduling were on small graphs to allow solving the scheduling problem optimally for both cases, and to allow illustrating the benefits of ANC over DNC. In this section, the performance for both scheduling schemes is compared and evaluated on relatively larger graphs. To perform the comparison, a simple greedy algorithm is proposed as shown in Algorithm 4.2.

The algorithm searches the unscheduled sources, and starts by selecting the source node with least degree to reduce conflicts with other sources as much as possible. The second step is to exclude any other source that, if scheduled in this time slot, will make the schedule infeasible. For example, in a DNC schedule, any source that have neighbors in L that are also neighbors of nodes in set S will be excluded. Such sources are put in the set of forbidden sources F. This process is repeated until no more sources can be added to the current time slot which indicates that additional time slots are needed if $|S| < |U|$.

The length of the schedule depends on network density which directly affects the node degree. Therefore, Algorithm 4.2 was executed using four different values for the degree of nodes in U, while varying the number of nodes in U from 2 to 20 and satisfying the connectivity condition of Chap. 2. Specifically, the degree D of nodes in U took the values 2, 3, 5, and 10. The results are shown in Figs. 4.13, 4.14, 4.15, and 4.16 for $D = 2$, $D = 3$, $D = 5$, and $D = 10$ respectively. As can be seen from the figures, the number of needed time slots increases as either the degree

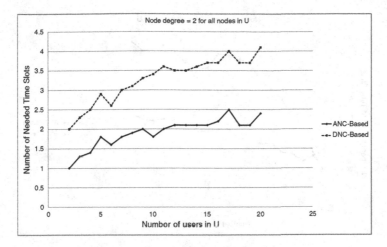

Fig. 4.13 Comparison between ANC and DNC. All U nodes have degree 2

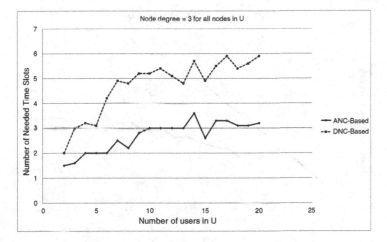

Fig. 4.14 Comparison between ANC and DNC. All U nodes have degree 3

of sources or their number increases. This is expected since increasing any of these parameters results in more conflicts. To gain more insight, the gain for all configurations is plotted in Fig. 4.17. When $|U| \leq 5$, the largest ANC gain is for the configuration with smallest degree. This is because when the number of nodes is small, having a small degree will help in satisfying the second feasibility condition for ANC-based scheduling, and thus, more sources can be grouped together. However, when $|U| \geq 12$, the best gain is for the configuration with the largest degree. This is because, if the number of nodes is small and the node degree is large, then the performance of DNC will be very close or similar to that of ANC. But, as the number of nodes increases, the performance of DNC will get worse faster than ANC because the DNC feasibility conditions are more strict, and thus, ANC

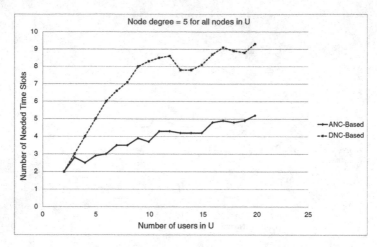

Fig. 4.15 Comparison between ANC and DNC. All U nodes have degree 5

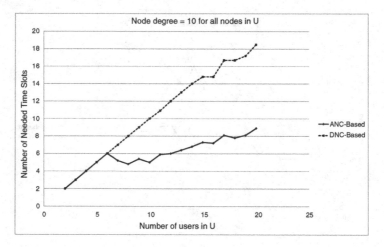

Fig. 4.16 Comparison between ANC and DNC. All U nodes have degree 10

starts to gain performance over DNC. Finally, note that although it is shown that ANC can outperform DNC by a factor of n theoretically, it is clear that on randomly generated graphs the actual gain of ANC is around 2 at best.

4.5 Chapter Summary

The problem of scheduling the transmissions of the source node in U to the coding nodes in L is studied. The scheduling problem was considered for both digital network coding (DNC) and analog network coding (ANC). It was shown that

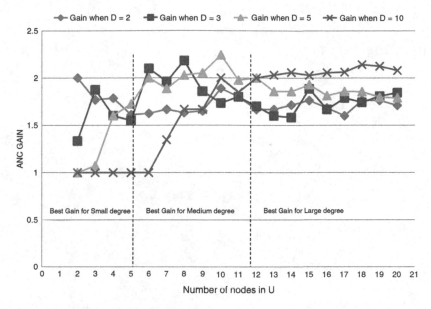

Fig. 4.17 ANC gain comparison for different node degree

applying ANC reduces the number of needed time slots in a schedule. In addition, some cases were studied in which the optimal ANC-based schedule can be found efficiently in polynomial time. Moreover, it was shown that using ANC can reduce the number of required time slots by at most a factor of n compared to using DNC, where n is the number of source nodes. This however, is highly unlikely to happen in practice. Both scheduling schemes were compared and evaluated on randomly generated graphs, and it was shown that the gain from using ANC is usually around 2.

References

1. S. Katti, S. Gollakota, and D. Katabi. Embracing wireless interference: Analog network coding. In the proceedings of SIGCOMM 2007.
2. R. Diestel. Graph theory (electronic edition 2005). Springer-Verlag Heidelberg, New York 1997, 2000, 2005.

Chapter 5
Conclusions

In WSNs, sensor nodes may fail to operate due to a number of reasons. Failures may also happen due to severe wireless communication link impairments, which affects data delivery between sensor nodes. For many applications, WSNs are required to remain operational and provide resilience against such failures. This brief deals with the problem of circumventing the effects of such failures, and of providing resilient operation.

The brief focused on resilient strategies using the technique of network coding. After an introduction to the WSNs resilience problem in Chap. 1, the technique of network coding was introduced in the same chapter. The basics and benefits of network coding were introduced, and examples illustrating such benefits, including using network coding to provide protection were introduced. The purpose of using network coding is to be able to provide resilience while achieving two highly desired objectives, namely, agility in recovery from failures, and minimality of resources used for providing this recovery. Since WSNs employ many-to-one, or convergecast, service modes, Chap. 2 introduced the theory for providing network coding-based protection for many-to-one flows in general, including the necessary and sufficient conditions for achieving this protection. The developments started by considering the basic case of a single node failure, and then was extended to the general case of multiple failures. Moreover, general network topologies, including the case in which the min-cut was not sufficient to protect all source nodes simultaneously was considered. Coding is an important part of network coding-based protection, which includes the locations of coding, and the generation of the codes. Since this problem is a hard problem, heuristic algorithms for solving the coding problem, using coefficients from the binary field, were developed.

The developments in Chap. 2 were based on the assumption of a centralized protection strategy. This requires global information collection about the WSN. To deal with large WSNs, or with WSNs in which the sensor nodes move, either intentionally through mobilizers, or unintentionally because of physical conditions like mud slides, Chap. 3 dealt with the development of distributed strategies for

© The Author(s) 2015
O.M. Al-Kofahi, A.E. Kamal, *Resilient Wireless Sensor Networks*, SpringerBriefs
in Electrical and Computer Engineering, DOI 10.1007/978-3-319-23965-1_5

applying network coding to achieve resilience. Such strategies do not require maintaining global information about the WSN. This approach tolerates both node and link failures. It is also resilient against data packet losses. Coding strategies were also developed for this strategy.

The most widely assumed network coding strategy is digital network coding, in which packet combinations are formed in the electronic domain. This requires packet reception, demodulation, decoding, and then combining, followed by coding, modulation and transmission. Another approach is performing network coding in the electromagnetic domain, in which packet interference results in the formation of packet combinations. This can result in reducing the number of transmissions and retransmissions. Two approaches for this coding, namely, physical layer network coding network coding have been introduced in the literature. While the implementation of both strategies are different, their effect is the same. Therefore, in Chap. 4 we considered the problem of packet transmission scheduling under both digital network coding and analog network coding, and introduced scheduling algorithms under goth techniques. Performance comparisons showed the advantages of using analog network coding over digital network coding in reducing the number of transmissions, hence increasing the system throughput.

Printed in the United States
By Bookmasters